T0190480

Use R!

Use R!

This series of inexpensive and focused books on R will publish shorter books aimed at practitioners. Books can discuss the use of R in a particular subject area (e.g., epidemiology, econometrics, psychometrics) or as it relates to statistical topics (e.g., missing data, longitudinal data). In most cases, books will combine LaTeX and R so that the code for figures and tables can be put on a website. Authors should assume a background as supplied by Dalgaard's Introductory Statistics with R or other introductory books so that each book does not repeat basic material.

More information about this series at http://www.springer.com/series/6991

Paulo Cortez

Modern Optimization with R

Second Edition

Paulo Cortez
Department of Information Systems
University of Minho
Guimarães, Portugal

ISSN 2197-5736 ISSN 2197-5744 (electronic)
Use R!
ISBN 978-3-030-72818-2 ISBN 978-3-030-72819-9 (eBook)
https://doi.org/10.1007/978-3-030-72819-9

This Springer imprint is published by the registered company Springer Nature Switzerland AG
The registered company address is: Gewerbestrasse 11, 6330 Cham, Switzerland

Preface

At present, we are in the age of data, where multiple individual and organizational activities and processes generate big data that can be processed using information technology (IT). We are also in a fast changing world. Due to several factors, such as globalization and technological improvements, both individuals and organizations are pressured for improving their efficiency, reducing costs, and making better informed decisions. This is where optimization methods, supported by computational tools, can play a key role.

Optimization is about minimizing or maximizing one or more goals and it is useful in several domains, including agriculture, banking, control, engineering, finance, management, marketing, production, and science. Examples of real-world applications include the optimization of construction works, financial portfolios, marketing campaigns, and water management in agriculture, just to name a few.

Modern optimization, also known as metaheuristics, is related with general-purpose solvers based on computational methods that use few domain knowledge from the addressed task, iteratively improving an initial solution (or population of solutions) to optimize a problem. Modern optimization is particularly useful for solving complex problems for which no specialized optimization algorithm has been developed, such as problems with discontinuities, dynamic changes, multiple objectives, and hard and soft restrictions, which cannot be handled easily by classical operational research methods.

Although modern optimization often incorporates random processes within their search engines, the overall optimization procedure tends to be much better than pure random (Monte Carlo) search. Several of these methods are naturally inspired. Examples of popular modern methods that are discussed in this book are simulated annealing, tabu search, genetic algorithms, genetic programming, grammatical evolution, NSGA II (multi-objective optimization), differential evolution, particle swarm, and ant colony optimization.

R is a free, open-source, and multiple platform tool (e.g., *Windows*, *macOS*, *Linux*) that was specifically developed for statistical analysis. Currently, there is an increasing interest in using R to perform an intelligent data analysis. In effect, the R community is very active and new packages are being continuously created. For example, there are currently more than 16,170 packages available at the Comprehensive R Archive Network (CRAN) (http://www.r-project.org/), which enhance the tool capabilities. In particular, several of these packages implement modern optimization methods.

There are several books that discuss either modern optimization methods or the R tool. However, within the author's knowledge, there is no book that integrates both subjects and under a practical point of view, with several application R code examples that can be easily tested by the readers. Hence, the goal of this book is to gather in a single document (self-contained) the most relevant concepts related with modern optimization methods, showing how such concepts and methods can be implemented using the R tool. It should be noted that some of the explored modern optimization packages have a minimal documentation (e.g., with no vignettes or very short help pages). Thus, this book can be used as complementary source for better understanding how to use these packages in practice. Moreover, some metaheuristics (e.g., simulated annealing, genetic algorithms, differential evolution) are implemented in distinct packages. In such cases, the book addresses all these packages, showing their similarities and differences, thus providing a broader (and hopefully not biased) view for the readers.

This book addresses several target audience groups. Given that the R tool is free, this book can be easily adopted in several bachelor's or master's level courses in areas such as Artificial Intelligence, operations research, data science, decision support, business intelligence, business analytics soft computing, or evolutionary computation. Thus, this book should be appealing for bachelor's or master's students in computer science, information technology, or related areas (e.g., engineering, science, or management). The book should also be of interest for two types of practitioners: R users interested in applying modern optimization methods and non R expert data analysts or optimization practitioners that want to test the R capabilities for optimizing real-world tasks.

First Edition Feedback

The first edition of this book was published in 2014 and it included a total of 188 pages. The book edition covered a publication gap, since there were no written books about the usage of metaheuristics with the R tool. Indeed, after publication, the book received positive feedback (http://www3.dsi.uminho.pt/pcortez/mor/), including two book reviews:

- "The author provides valuable comments about the pros and cons of various optimization methods. ... This book makes a good contribution in to the literature of modern optimization. It is well written and structured." – Diego Ruiz, Journal of Statistical Software, 2016 (https://www.jstatsoft.org/article/view/v070b03/v70b03.pdf); and

- "...the positive characteristics of Cortez's Modern Optimization with R make this book a must-have for graduate students and any researchers willing to delve into new algorithms to solve their optimisation tasks. I was pleasantly impressed by the book's organisation and clarity, by its comprehensive and coherent topic coverage, and by its many outstanding numerical illustrations using R." – Matthieu Vignes, Australian & New Zealand Journal of Statistics, 2017 (https://onlinelibrary.wiley.com/doi/epdf/10.1111/anzs.12186).

Moreover, the book was adopted as a textbook by several bachelor's and master's level university courses, such as: "Discrete and Continuous Optimization" (University of California San Diego, USA); "Operations Research" (University of Loisville, USA); "Stochastic Optimization" (North Carolina State University, USA); "Advanced Optimization" (Warsaw School of Economics, Poland); "Machine Learning for Finance" (Barcelona Graduate School of Economics, Spain); and "Optimization Techniques" (National Institute of Technology Delhi, India). Furthermore, the book has received 65 Google scholar (https://scholar.google.com/) citations. In 2019, the book was among the top 25% most downloaded e-books of its respective collection. The book is also mentioned in the CRAN task view for Optimization and Mathematical Programming (https://cran.r-project.org/web/views/Optimization.html).

Updated and Revised Second Edition

This second edition consists of an updated version that includes 270 pages. While the core structure and material is essentially the same, the whole text was proofread and several portions were rewritten to reflect the teaching and research experience of the author. Moreover, several of the R code demonstrations were substantially updated. In effect, the first edition was written in 2013, when there were around 5,800 R packages. Since then, this number as almost tripled, with several of these newer packages implementing metaheuristics. Examples of packages that are now explored in this second edition include: **DEoptimR**, **GenSA**, **GA**, **MaOEA**, **NMOF**, **ecr**, **evoper**, **gramEvol**, **irace**, **mcga**, and **psotim**. New material was also added, such as: limitations of metaheuristics, interfacing with the Python language, creating interactive Web applications via the **shiny** package, a stronger emphasis on parallel computing (with several examples, including the island model for genetic algorithms), iterated racing, ant colony optimization, evolutionary breeding and selection operators, grammatical evolution, and several multi-objective algorithms (e.g., SMS-EMOA, AS-EMOA, NSGA-III).

The second edition code was written using the version 4.0.0 of the R tool, while the first edition was produced using the older version 3.0.0. Before updating the R demonstrations, all first edition R code examples (written in 2013) were re-executed using the more recent R version (4.0.0). In some cases, the executions used newer versions of the previously adopted packages (e.g., **genalg** latest version was written in 2015). While 7 years have passed (since 2013), it was interesting to notice that the code examples executed well in almost all cases. This attests the robustness of the first edition code in the R tool technological evolution. The author expects that a similar phenomenon occurs with the second edition code. However, if needed and whenever possible, code updates will be provided at: https://www.springer.com/us/book/9783030728182.

How to Read This Book

This book is organized as follows:

Chapter 1 introduces the motivation for modern optimization methods and why the R tool should be used to explore such methods. Also, this chapter discusses key modern optimization topics, namely the representation of a solution, the evaluation function, and how to handle constraints. Then, an overall view of modern optimization methods is presented, followed by a discussion of their limitations and criticism. This chapter ends with the description of the optimization tasks that are used for tutorial purposes in the next chapters.

Chapter 2 presents basic concepts about the R tool. Then, more advanced features are discussed, including command line execution, parallel computing, interfacing with other computer languages, and interactive Web applications. This chapter is particularly addressed to non R experts, including the necessary knowledge that is required to understand and test the book examples. R experts may skip this chapter.

Chapter 3 is about how blind search can be implemented in R. This chapter details in particular two full-blind search implementations, two grid search approaches (standard and nested), and a Monte Carlo (random) search.

Chapter 4 discusses local search methods, namely hill climbing (pure and steepest ascent and stochastic variants), simulated annealing, and tabu search. Finally, it shows how modern optimization methods (including local search ones) can be tuned in terms of their internal parameters by using an iterated racing.

Chapter 5 presents population-based search methods, namely genetic and evolutionary algorithms, differential evolution, particle swarm optimization, ant colony optimization, estimation of distribution algorithm, genetic programming, and grammatical evolution. The chapter also presents examples of how to compare population based methods, how to handle constrains, and how to run population-based methods in parallel.

Chapter 6 is dedicated to multi-objective optimization. First, three demonstrative multi-objective tasks are presented. Then, three main multi-objective approaches are discussed and demonstrated: weighted-formula, lexicographic, and Pareto (e.g., NSGA-II and NSGA-III, SMS-EMOA, AS-EMOA).

Chapter 7 presents real-world applications of previously discussed modern optimization methods: traveling salesman problem, time series forecasting, and wine quality classification.

Each chapter starts with an introduction, followed by several chapter-topic related sections and ends with an R command summary and exercises sections. Throughout the book, several examples of R code are shown. The code was run using a 64 bit R (version 4.0.0) under on a *macOS* laptop. Nevertheless, these examples should be easily reproduced by the readers on other systems, possibly resulting in slight numerical (32 bit version) or graphical differences for the deterministic examples. Also, given that a portion of the discussed methods are stochastic, it is natural that different executions of the same code and under the same system will lead to (slight) differences in the results.

It is particularly recommended that students should execute the R code and try to solve the proposed exercises. Examples of solutions for proposed exercises are presented at the end of this book. All these code files and data examples are available at: https://www.springer.com/us/book/9783030728182.

Production

Several contents of this book were taught by the author in the last 12 years in distinct course units of master's and doctoral programs. At the master's level, it included the courses "adaptive business intelligence" (Master of Engineering and Management of Information Systems, University of Minho, Portugal) and "business intelligence" (Master of Information System Management, Lisbon University Institute, Portugal). The doctoral course was "adaptive business intelligence" (Doctoral Program in Computer Science, Universities of Minho, Aveiro and Porto, Portugal). Also, some of the older material was lectured at a tutorial given in the European Simulation and Modelling Conference (ESM 2011), held at Guimarães.

This book was written in LaTeX, using the `vim` editor (http://vim.org) and its US English spell checker. Most figures were made in R, while some of the figures were designed using `xfig` (http://www.xfig.org), an open source drawing tool.

Guimarães, Portugal Paulo Cortez

Contents

List of Figures

List of Algorithms

Chapter 1
Introduction

1.1 Motivation

Advances in Information Technology (IT) and Artificial Intelligence (AI) are highly impacting the world. Since the 1970s, and following the Moore's law, the number of transistors in computer processors has doubled every two years, resulting in more computational power at a reasonable price. And it is estimated that the amount of data storage doubles at a higher rate due to many big data sources, such as social media, Internet-of-Things (IoT), industry 4.0, and smart cities. Following this computational power growth and data explosion, there has been an increased usage of intelligent data analysis tools, proposed under different perspectives (Cortez and Santos, 2015): machine learning, since the 1960s (including deep learning, since the 2010s); metaheuristics, since the 1960s; decision support systems, since the 1970s; analytics, data mining, and business intelligence, since the 1990s; and big data, data science, and business analytics, since the 2000s.

Moreover, nowadays there is a pressure in organizations and individual users to increase efficiency and reduce costs (Turban et al., 2010). Rather than taking decisions based on human experience and intuition, there is a growing trend for adopting computational tools, based on optimization methods, to analyze real-world data in order to make better informed decisions. The famous Garner group proposed in 2012 a four-level business analytics model (Fig. 1.1) (Koch, 2015). The model assumes an increasing value and difficulty in the business analyses types, going from information-based (more descriptive) to optimization-based (more actionable knowledge). The fourth level, prescriptive analytics, is expected to produce the highest impact by answering to the question: "how can we make it happen?". Several techniques can be used to perform prescriptive analyses, which include optimization and other methods (e.g., what-if scenario analysis, simulation).

P. Cortez, *Modern Optimization with R*, Use R!,
https://doi.org/10.1007/978-3-030-72819-9_1

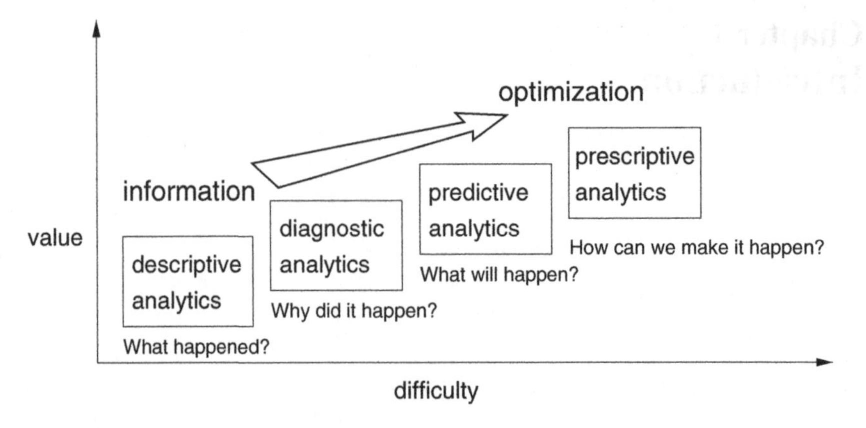

Fig. 1.1 Gartner four-level analytics model. (Adapted from Koch, 2015)

A vast number of real-world (often complex) tasks can be viewed as an **optimization** problem, where the goal is to **minimize or maximize a given goal**. In effect, optimization is quite useful in distinct application domains, such as Agriculture, Banking, Control, Engineering, Finance, Marketing, Production, Tourism, and Science. Optimization is a core topic of the Operations Research field, which developed several classical mathematical analysis techniques, such as linear programming (proposed in the 1940s) and branch and bound (developed in the 1960s) (Schrijver, 1998). More recently, in the last decades, there has been an emergence of new AI optimization algorithms, often termed as "modern optimization" (Michalewicz et al., 2006; López-Ibáñez et al., 2016), "modern heuristics" (Michalewicz and Fogel, 2004), or "metaheuristics" (Luke, 2015). It should be noted that the term metaheuristics can mention both the problem-independent algorithmic framework and the specific algorithms built using its principles (Sörensen, 2015). In this book, we adopt the first term, modern optimization, to describe these AI algorithms.

In contrast with classical methods, modern optimization methods are general-purpose solvers, i.e., applicable to a wide range of distinct problems, since little domain knowledge is required. For instance, the optimization problem does not need to be differentiable, which is required by classical methods such as gradient descent. There are only two main issues that need to be specified by the user when adopting modern heuristic methods (Michalewicz et al., 2006): the representation of the solution, which defines the search space and its size; and the evaluation function, which defines how good a particular solution is, allowing to compare different solutions. In particular, modern methods are useful for solving complex problems for which no specialized optimization algorithm has been developed (Luke, 2015). For instance, problems with discontinuities, dynamic changes, multiple objective, or hard and soft restrictions, which are more difficult to be handled by classical methods (Michalewicz et al., 2006). Also in contrast with classical methods, modern optimization does not warranty that the optimal solution is always found. However, often modern methods achieve high quality solutions with a much more reasonable use of computational resources (e.g., memory and processing effort) (Michalewicz and Fogel, 2004).

There are a vast number of successful real-world applications based on modern optimization methods. Examples studied by the author of this book include (among others): sitting guests at a wedding party (Rocha et al., 2001); optimization of data mining classification and regression models (Rocha et al., 2007); improvement of quality of service levels in computer networks (Rocha et al., 2011); allocation of compaction equipment for earthworks (Parente et al., 2015); assisting the design of online news (Fernandes et al., 2015); predicting user conversion responses to mobile marketing campaigns (Pereira et al., 2019); estimation of time series prediction intervals (Cortez et al., 2020); and rebalancing of shared electric scooters (Fernandes et al., 2020).

1.2 Why R?

The R tool (R Core Team, 2020) is an open source, high-level matrix programming language for statistical and data analysis. The tool runs on multiple platforms, including *Windows*, *MacOS*, *Linux*, *FreeBSD*, and other UNIX systems. R is an interpreted language, meaning that the user gets an immediate response of the tool, without the need of program compilation. The most common usage of R is under a console command interface, which often requires a higher learning curve from the user when compared with other more graphical user interface tools. However, after mastering the R environment, the user achieves a better understanding of what is being executed and higher control when compared with black box graphical interface based products. Moreover, R can be used to automate reports and it can be integrated with cloud computing environments and database systems (Quantargo, 2020).

The R base distribution includes a large variety of statistical techniques (e.g., distribution functions, statistical tests), which can be useful for inclusion in modern optimization methods and to analyze their results. Moreover, the tool is highly extensible by creating packages. The R community is very active and new packages are being continuously created, with more than 16,170 packages available at the Comprehensive R Archive Network (CRAN): http://www.r-project.org/. By installing these packages, users get access to new features, such as: data mining/machine-learning algorithms; simulation and visualization techniques; and also modern optimization methods. New algorithms tend to be quickly implemented in R; thus, this tool can be viewed as worldwide gateway for sharing computational algorithms (Cortez, 2010). It is difficult to know the real number of R users. In 2012, it was estimated that there were 2 million R users (Vance, 2009). In Muenchen (2019), R was ranked at fifth place in terms of popularity within data science jobs. The IEEE Spectrum magazine also ranks R as the fifth programming language (Cass, 2019), which is an impressive result since this language was specifically developed for data analysis and thus it is not as generic as other computer programming languages (e.g., Java, C, Python). Examples of companies using R include (Fay, 2020; Quantargo, 2020): Airbnb, Allianz, Amazon, Booking, eBay, Facebook, Google, Microsoft, Netflix, Oracle, Pfizer, Twitter, and Uber. A useful advantage of using R is that it is possible to execute quite distinct computational tasks under the same tool, such as combining optimization with statistical analysis, visualization, simulation, and machine learning (see Sect. 7.4 for an example that optimizes machine-learning models).

To facilitate the usage of packages, given that a large number is available, several R packages are organized into CRAN task views. The Optimization and Mathematical Programming view is located at http://cran.r-project.org/web/views/Optimization.html and includes more than 130 packages. In this book, we explore several of these CRAN view packages (and others) related with modern optimization methods. Whenever possible, the book opted to select stable packages that did not crash and did not require the installation of other computational tools (e.g., Python libraries). Moreover, some modern optimization methods have distinct (and independent) package implementations. For example, there are at least five packages that implement evolutionary algorithms (**GA**, **NMOF**, **ecr**, **genalg**, and **mcga**). As explained in the preface, in such cases and whenever possible, the book explores all independent packages, detailing their similarities and differences.

1.3 Representation of a Solution

A major decision when using modern optimization methods is related with how to represent a possible solution (Michalewicz et al., 2006). Such decision sets the search space and its size, thus producing a high impact on how new solutions are searched. To represent a solution, there are several possibilities. Binary, integer, character, and real value and ordered vectors, matrices, trees, and virtually any computer based representation form (e.g., computer program) can be used to encode solutions. A given representation might include a mix of different encodings (e.g., binary and real values). Also, a representation might be of fixed (e.g., fixed binary vectors) or of variable length (e.g., trees).

Historically, some of these representation types are attached with specific optimization methods. For instance, binary encodings are the basis of genetic algorithms (Holland, 1975). Tabu search was designed to work on discrete spaces (Glover and Laguna, 1998). Real value encodings are adopted by several evolutionary algorithms (e.g., evolution strategies) (Bäck and Schwefel, 1993), differential evolution, and particle swarm optimization. Tree structures are optimized using genetic programming (Banzhaf et al., 1998) and computer programs with a syntax can be evolved using grammatical evolution (Ryan et al., 1998). It should be noted that often these optimization methods can be adapted to other representations. For instance, a novel particle swarm was proposed for discrete optimization in Chen et al. (2010).

In what concerns this book, the representations adopted are restricted by the implementations available in the explored R tool packages, which mainly adopt binary or real values. Thus, there will be more focus on these representation types. When the selected representation is not aligned with the method implementation, a simple solution is to use a converter code at the beginning of the evaluation function. For instance, the **tabuSearch** package only works with binary implementations; thus a binary to integer code was added in Sect. 4.4 to optimize the bag prices task. Similarly, several population based method implementations (e.g., **pso**) only work with real values and thus a round function was used in the bag prices evaluation function to convert the solutions into the desired integer format (Sect. 5.7). The same strategy was employed in Sect. 6.5, where the real solution values are rounded into a binary (`0` or `1`) values, allowing to apply the NSGAII algorithm for a multi-objective binary task. Given the importance of this topic, some simple representation examples are further discussed in this section. Section 7.2 demonstrates an ordered vector (permutation based) representation. More details about other representations, including their algorithmic adjustments, can be found in Luke (2015).

The representation of the solution should be as "close" as possible to the task being solved. Also, the search space should be as small as possible. For instance, consider the task of *defining a team with any of the members* from the set {Anna, Beatrice, Charles, David, Emily}. If a binary encoding is used, this task is best represented by using a dimension of D =5 bits: $(b_1, b_2, b_3, b_4, b_5)$, where each $b_i \in \{0, 1\}$ bit defines if the element is present in the team (if $b_i = 1$ it means "true"). Under this encoding, $(1, 0, 1, 0, 0)$ represents the set {Anna, Charles}. In this case, the search space contains $2^5 = 32$ possible solutions. However, if the task is changed to *select one of the members*, then a better representation is a single integer n ($D = 1$), where n is a natural number that could assume the following encoding: $1 \rightarrow$ Anna, $2 \rightarrow$ Beatrice, $3 \rightarrow$ Charles, $4 \rightarrow$ David, and $5 \rightarrow$ Emily. The second representation has a smaller search space, with just 5 possibilities. It is a more natural representation for the second optimization task than a five bit representation, because all 5 solutions are valid, while the binary encoding would result in 27 infeasible solutions, such as $(0, 1, 0, 1, 0)$ or $(1, 0, 1, 1, 0)$. Another possibility would be to use a $D = 3$ bit binary encoding, with the base 2 representation of the integer encoding, such as: $(0, 0, 0) \rightarrow 1$ (Anna), $(0, 0, 1) \rightarrow 2$ (Beatrice), and so on. However, this $D = 3$ bit representation would still have 3 infeasible solutions, such as $(1, 0, 1) \rightarrow 6$ or $(1, 1, 0) \rightarrow 7$. Both binary and integer representations are widely used in discrete optimization tasks, where the variables only take on discrete values.

Another representation example consists in the optimization of 20 continuous numeric parameters. In this case, rather than using binary or integer encodings, the solutions are often better represented by using a vector with a dimension of $D = 20$ real values. This last example consists of a numerical or continuous optimization task, which typically involves bounded real values, with lower and upper limits (e.g., $\in [0, 1]$), to set the search space.

phase 1			phase 2			phase 3		
c_1	c_2	c_3	c_1	c_2	c_1	c_1	c_2	c_3
2	1	1	1	1	3	3	0	3

Fig. 1.2 Example of a compactor allocation with 3 compactors and 3 construction fronts. (Adapted from Parente et al., 2015)

In complex real-world tasks, the representation of the solution can be nontrivial. For example, earthworks use a wide range of equipment (e.g., excavators, trucks, spreaders, and compactors) that need to be assigned to multiple construction fronts. In Parente et al. (2015), domain knowledge was used to simplify the representation of solutions: soil compaction (the last task of the earthworks) should only be used if there is material to be compacted. Thus, only compactors were allocated by the optimization method, with the remaining equipment (e.g., trucks, excavators) being automatically distributed according to the compactor allocation needs. The solutions were represented by using an integer encoding n_i with a dimension of $D = C \times F$, where C denotes the number of compactors, F the number of construction fronts and $i \in \{1, \ldots, C \times F\}$. The adopted representation uses a sequence of up to F construction phases, assuring that all fronts were compacted. During each phase, it was assumed that a compactor works in only one front; thus $n_i \in \{0, 1, \ldots, F\}$, where 0 means that the compactor is not used. An example of a possible solution when using three compactors ($C = 3$, from the set $\{c_1, c_2, c_3\}$) and three fronts ($F = 3$) is shown in Fig. 1.2. In this example, the compactor c_2 is used in front 1 during the first two phases and then it is idle in phase 3.

1.4 Evaluation Function

Another important decision for handling optimization tasks is the definition of the evaluation function, which should translate the desired goal (or goals) to be maximized or minimized. Such function allows to compare different solutions, by providing either a rank (ordinal evaluation function) or a quality measure score (numeric function) (Michalewicz et al., 2006). When considering numeric functions, the shape can be convex or non-convex, with several local minima/maxima (Fig. 1.3). Convex tasks are much easier to solve and there are specialized algorithms (e.g., least squares, linear programming) that are quite effective for handing such problems (Boyd and Vandenberghe, 2004). However, many practical problems are non-convex, often including noisy or complex function landscapes, with discontinuities. Optimization problems can even be dynamic, changing through time. For all these complex problems, an interesting alternative is to use modern optimization algorithms that only search a subset of the search space but tend to achieve near optimum solutions in a reasonable time.

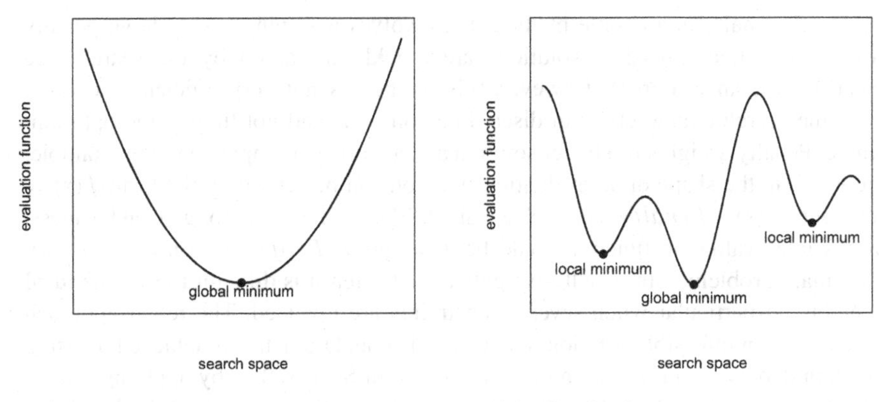

Fig. 1.3 Example of a convex (left) and non-convex (right) function landscapes

By default, some implementations of optimization methods only perform a minimization of the numerical evaluation function. In such cases, a simple approach is to transform the maximization function $\max(f(s))$ into the equivalent minimization task $-\min(f'(s))$, by adopting $f'(s) = K - f(s)$, where s denotes the solution and K is a constant value. Often, the default $K = 0$ value can be used, resulting in $f'(s) = -f(s)$. However, if there are constraints in the function codomain (e.g., f' needs to produce only positive values), then another K value can be adopted (often set using domain knowledge).

In several application fields (e.g., Control, Engineering, Finance) there are two or more goals that need to be optimized. Often, these goals conflict and trade-offs need to be set, since optimizing solutions under a single objective can lead to unacceptable outcomes in terms of the remaining goals. In such cases, a much better approach is to adopt a multi-objective optimization. In most of this book, we devote more attention to single response evaluation functions, since multi-objective optimization is discussed in a separate chapter (Chap. 6).

1.5 Constraints

There are two main types of constraints (Michalewicz, 2008): hard and soft. Hard constraints cannot be violated and are due to factors such as laws or physical restrictions. Soft constraints are related with other (often non-priority) user goals, such as increasing production efficiency while reducing environmental costs.

Soft restrictions can be handled by adopting a multi-objective approach (Chap. 6), while hard constraints may originate infeasible solutions that need to be treated by the optimization procedure. To deal with infeasible solutions, several methods can be adopted (Michalewicz et al., 2006): death-penalty, penalty-weights, repair, and only generate feasible solutions.

Death-penalty is a simple method that involves assigning a very large penalty value, such that infeasible solutions are quickly discarded by the search (see Sect. 4.4 for an example). However, this method is not very efficient and often puts the search engine effort in discarding solutions and not finding the optimum value. Penalty-weights is a better solution and also easy to implement. For example, quite often, the shape of an evaluation function can be set within the form $f(s) = Objective(s) - Penalty(s)$ (Rocha et al., 2001). For instance, for a given business, a possible evaluation function could be $f = w_1 \times Profit(s) - w_2 \times Cost(s)$. The main problem with penalty-weights is that often it is difficult to find the ideal weights, in particular when several constraints are involved. The repair approach transforms an infeasible solution into a feasible one. Often, this is achieved by using domain dependent information (such as shown in Sect. 5.2) or by applying a local search (e.g., looking for a feasible solution in the solution space neighborhood, see Sect. 5.8). Finally, the approaches that only generate feasible solutions are based on decoders and special operators. Decoders work only in a feasible search space, by adopting an indirect representation, while special operators use domain knowledge to create new solutions from previous ones.

1.6 Optimization Methods

There are different dimensions that can be used to classify optimization methods. Three factors of analysis are adopted in this book: the type of guided search; if the search is deterministic or stochastic based; and if the method is inspired by physical or biological processes.

The type of search can be blind or guided. The former assumes that a previous search does not affect the next search, while the latter uses previous searches to guide the current search. Modern methods use a guided search that is often subdivided into two main categories: single-state, which typically searches within the neighborhood of an initial solution, and population based, which uses a pool of solutions and typically performs a global search. In most practical problems, with high-dimensional search spaces, pure blind search is not feasible, requiring too much computational effort. Single-stage or local search presents in general a much faster convergence, when compared with global search methods. However, if the evaluation landscape is too noisy or complex, with several local minima (e.g., right of Fig. 1.3), local methods can easily get stuck. In such cases, multiple runs, with different initial solutions, can be executed to improve convergence. Although population based algorithms tend to require more computation than local methods, they perform a simultaneous search in distinct regions of the search space, thus working better as global optimization methods.

Several modern methods employ some degree of randomness, thus belonging to the family of stochastic optimization methods, such as simulated annealing and genetic algorithms (Luke, 2015). Also, several of these methods are naturally inspired (e.g., genetic algorithms, particle swarm optimization) (Holland, 1975;

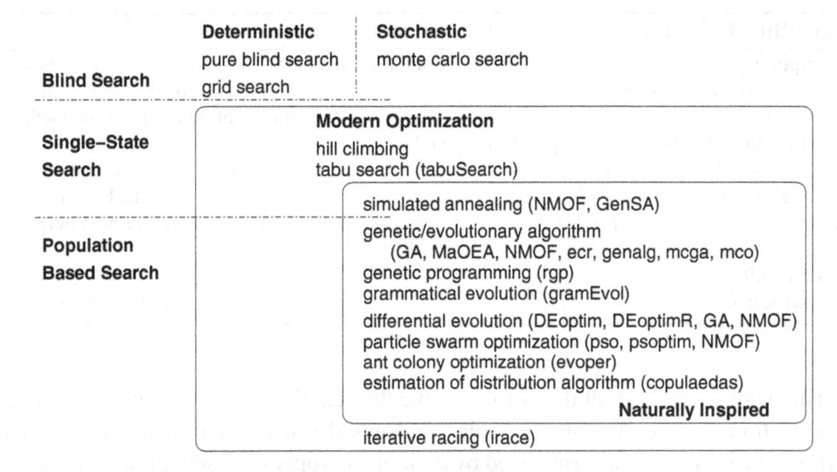

Fig. 1.4 Classification of the optimization methods presented in this book (related R packages are in brackets)

Eberhart et al., 2001). Figure 1.4 shows the full taxonomy of the optimization methods presented in this book (with respective R packages).

The distinct modern optimization methods share some common features. Algorithm 1 shows (in pseudo-code) a generic modern optimization method that is applicable to most methods discussed in this book. This global algorithm receives two inputs, the evaluation function (f) and a set of control parameters (C) that includes not only the method's internal parameters (e.g., initial temperature, population size) but also related with the representation of the solution (e.g., lower and upper bounds, representation type, and length). In all modern optimization methods, there is an initial setup followed by a main loop cycle that ends once the given termination criteria are met. Distinct criteria can be adopted (or even combined):

- Maximum computational measures—such as number of iterations, evaluation function calculations, and time elapsed;
- Target measures—such as to stop if best value is higher or equal to a given threshold;
- Convergence measures—such as the number of iterations without any improvement in the solution or average enhancement achieved in the last iterations;
- Distribution measures—such as measuring how the last tested solutions spread in the search space and stop if the dispersion is smaller than a threshold.

What distinguishes the methods is related with two main aspects. First, in each iteration there is a single-state (local based) or a population of solutions. Second, the way new solutions are created (function *change*) and used to guide in the search (function *select*). In the generic pseudo-code, the number of iterations (i) is also included as an input of the *change*, *best*, and *select* functions because it is assumed that the behavior of these functions can be dynamic, changing as the search method evolves.

Algorithm 1 Generic modern optimization method

1: **Inputs:** f, C ▷ f is the evaluation function, C includes control parameters
2: $S \leftarrow initialization(C)$ ▷ S is a solution or population
3: $i \leftarrow 0$ ▷ i is the number of iterations of the method
4: **while** not $termination\_criteria(S, f, C, i)$ **do**
5: $S' \leftarrow change(S, f, C, i)$ ▷ new solution or population
6: $B \leftarrow best(S, S', f, C, i)$ ▷ store the best solution
7: $S \leftarrow select(S, S', f, C, i)$ ▷ solution or population for next iteration
8: $i \leftarrow i + 1$
9: **end while**
10: **Output:** B ▷ the best solution

There are several possibilities to improve the search results when using modern optimization methods to solve complex real-world tasks. For instance, the search performance can be highly enhanced by defining a proper representation of the solution, evaluation function, and how constraints are handled (discussed in Sects. 1.3, 1.4, and 1.5). Another important aspect is selecting and adjusting the optimization method (discussed here and throughout this book). This might involve comparing distinct methods, tuning of their C control parameters (e.g., the temperature in a simulated annealing, Sect. 4.3), or even adjusting how a particular method works (e.g., the *select* and *change* functions from Algorithm 1). A simple example of the last option is the selection of the crossover and mutation operators (thus adjusting the *change* function) to be used by a genetic algorithm (Sect. 5.2). Moreover, the distinct search types can be combined. For instance, a two phase search can be set, where a global method is employed at a first step, to quickly identify interesting search space regions, and then, as the second step, the best solutions are improved by employing a local search method. Another alternative is to perform a tight integration of both approaches, such as under a Lamarckian evolution or Baldwin effect (Rocha et al., 2000). In both cases, within each cycle of the population based method, each new solution is used as the initial point of a local search and the evaluation function is computed over the improved solution. Lamarckian evolution replaces the population original solution by its improved value (Sect. 7.2 presents a Lamarckian evolution example), while the Baldwinian strategy keeps the original point (as set by the population based method). Finally, domain knowledge can also be used to set the initial search, to perform a local search or when handling constraints, such as exemplified in Sects. 5.8 and 7.2. However, using too much domain knowledge makes the implemented search approach too specific and less adaptable to other optimization tasks (as shown in Sect. 7.2).

1.7 Limitations and Criticism

Modern optimization methods have several limitations (some already mentioned in this chapter). First, they do not warranty that the optimum solution will always be found. If the search space is small (computationally feasible), and an optimum solution is really needed, then pure blind search is a better solution. Second, since little domain knowledge is used, metaheuristics can have a poor performance in specialized tasks when compared with operations research mathematical approaches. An example is shown in Sect. 7.2, where the specialized 2-opt method performs better than the generic evolutionary algorithm for the Traveling Salesman Problem. However, it should be noted that several specialized methods assume a simplified model of the world. Complex real-work optimization tasks often contain constraints, discontinuities, and dynamic changes, which are better handled by metaheuristics (Michalewicz et al., 2006). Third, the *No Free Lunch (NFL)* theorem states that search algorithms have a similar mean performance when tested over the set of all possible objective functions. In other words, there is no well performing universal metaheuristic. However, the NFL theorem is only applicable when approaching all search tasks, which includes a wide range of random functions. In practice, there are differences between algorithms and some metaheuristics work better on some types of problem classes (Mendes et al., 2004; Igel, 2014). Often, empirical evidence is needed to attest such performance. Thus, different optimization methods should be compared when approaching a new task. To be more robust, the comparison should include several runs (such as shown in Sects. 4.5 and 5.7), which requires computational effort. Fourth, there has been an explosion of "novel" metaheuristics. As argued by Sörensen (2015), too many natural or man-made metaphors have inspired the proposal of "new" optimization methods that tend to be a way of selling the existing ideas under a different name or metaphor. Examples of such metaphors include (Sörensen, 2015): colonialism—imperialist competitive algorithm; insects—fruit fly optimization algorithm; flow of water—intelligent water drops algorithm; birds—cuckoo search; and jazz—harmony search. By using new terminologies, these "novel" methods often obfuscate commonalities with the existing methods and are more difficult to be understood.

1.8 Demonstrative Problems

This section includes examples of simple optimization tasks that were selected for tutorial purposes, aiming to easily show the capabilities of the optimization methods. The selected demonstrative problems include 2 binary, 1 integer, and 2 real value tasks. More optimization tasks are presented and explored in Chaps. 6 (multi-optimization tasks) and 7 (real-world tasks).

The binary problems are termed here **sum of bits** and **max sin**. The former, also known as max ones, is a simple binary maximization "toy" problem, defined as (Luke, 2015):

$$f_{\text{sum of bits}}(\mathbf{x}) = \sum_{i=1}^{D} x_i \tag{1.1}$$

where $\mathbf{x} = (x_1, \ldots, x_D)$ is a boolean vector ($x_i \in \{0, 1\}$) with a dimension (or length) of D. The latter problem is another simple binary task, where the goal is to maximize (Eberhart and Shi, 2011):

$$\begin{array}{c} x' = \sum_{i=1}^{D} x_i 2^{i-1} \\ f_{\text{max sin}}(\mathbf{x}) = \sin\left(\pi \frac{x'}{2^D}\right) \end{array} \tag{1.2}$$

where x' is the integer representation of x.

A visualization for both binary problems is given in top of Fig. 1.5, assuming a dimension of $D = 8$. In the top left graph, x−axis denotes x', the integer representation of $\mathbf{x}$ for **sum of bits**. In the example, the optimum solution for the **sum of bits** is $\mathbf{x} = (1, 1, 1, 1, 1, 1, 1, 1)$ ($f(\mathbf{x}) = 8$), while the best solution for **max sin** is $\mathbf{x} = (1, 0, 0, 0, 0, 0, 0, 0)$, $x' = 128$ ($f(\mathbf{x}) = 1$).

The **bag prices** is an integer optimization task (proposed in this book) that mimics the decision of setting of prices for items produced at a bag manufacturer factory. The factory produces up to five ($D = 5$) different bags, with a unit cost of $\mathbf{u} = (\$30, \$25, \$20, \$15, \$10)$, where u_i is the cost for manufacturing product i. The production cost is $cost(x_i) = 100 + u_i \times sales(x_i)$ for the i-th bag type. The number of expected sales, which is what the factory will produce, is dependent on the product selling price ($\mathbf{x}$) and marketing effort ($\mathbf{m}$), according to the formula $sales(x_i) = round((1000/\ln(x_i + 200) - 141) \times m_i)$, where $round$ is the ordinary rounding function and $\mathbf{m} = (2.0, 1.75, 1.5, 1.25, 1.0)$. The manager at the factory needs to decide the selling prices for each bag (x_i, in \$), within the range \$1 to \$1000, in order to maximize the expected profit related with the next production cycle:

$$f_{\text{bag prices}} = \sum_{i=1}^{D} [x_i \times sales(x_i) - cost(x_i)] \tag{1.3}$$

The middle left graph of Fig. 1.5 plots the full search space for the first item of **bag prices** ($D = 1$), while the middle right plot shows a zoom near the optimum solution. As shown by the graphs, the profit function follows in general a global convex shape. However, close to the optimum ($x_1 = 414$, $f(x_1) = 11420$) point there are several local minima, under a "saw" shape that is more difficult to optimize. As shown in Chap. 3, the optimum solution for five different bags ($D = 5$) is $\mathbf{x} = c(414, 404, 408, 413, 395)$, which corresponds to an estimated profit of \$43899.

Turning to the real value tasks, two popular benchmarks are adopted, namely **sphere** and **rastrigin** (Tang et al., 2009), which are defined by:

$$\begin{array}{l} f_{\text{sphere}}(\mathbf{x}) = \sum_{i=1}^{D} x_i^2 \\ f_{\text{rastrigin}}(\mathbf{x}) = \sum_{i=1}^{D} (x_i^2 - 10\cos 2\pi x_i + 10) \end{array} \tag{1.4}$$

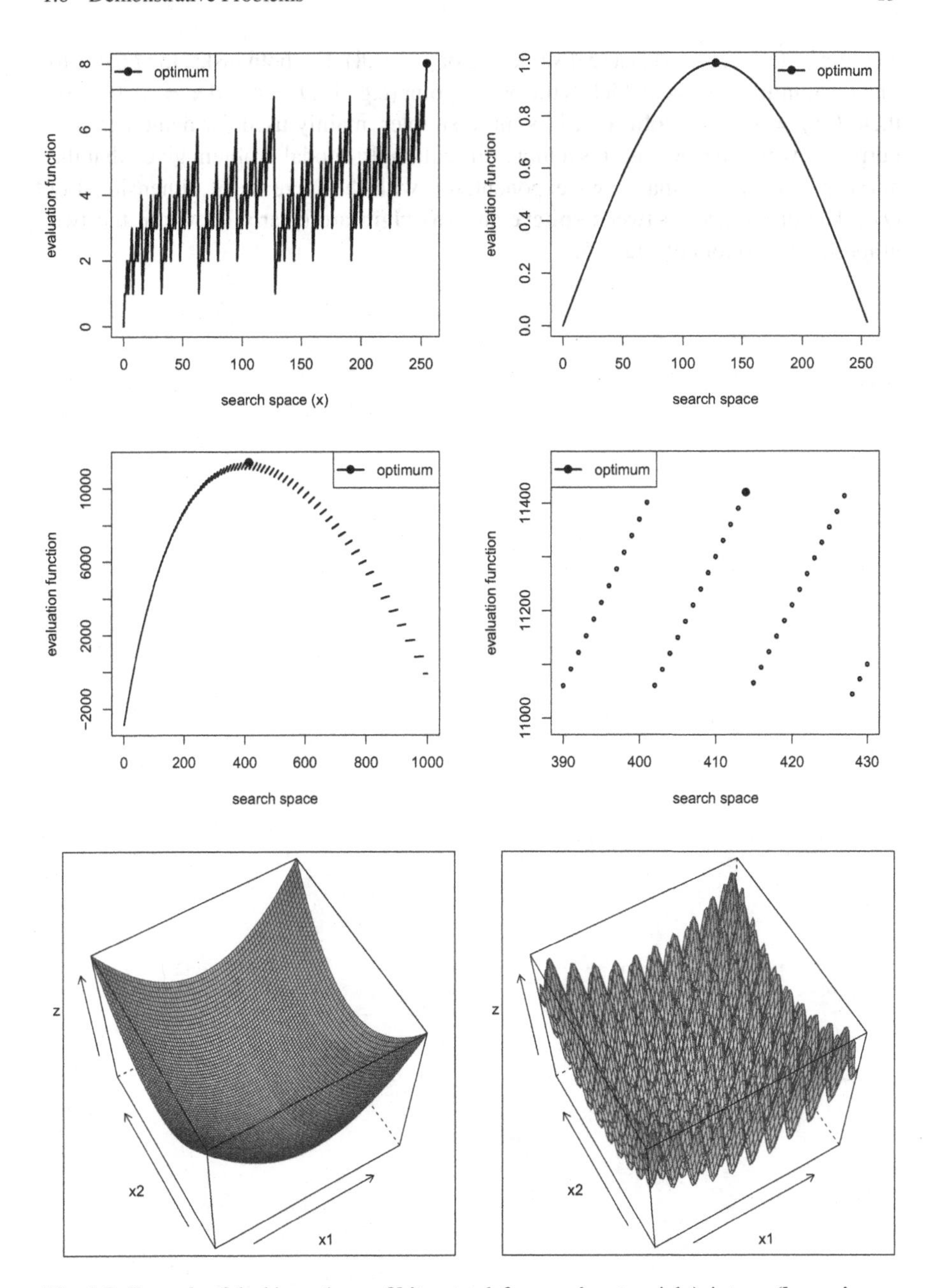

Fig. 1.5 Example of the binary (**sum of bits**—top left; **max sin**—top right), integer (**bag prices**—middle), and real value (**sphere**—bottom left; **rastrigin**—bottom right) task landscapes

where $\mathbf{x} = (x_1, \ldots, x_D)$ is a real value vector ($x_i \in \Re$). For both tasks, the goal is to find the minimum value, which is the origin point (e.g., if $D = 4$ and $\mathbf{x} = (0, 0, 0, 0)$ then $f(\mathbf{x}) = 0$). The **sphere** task is more simpler, mainly used for demonstration purposes, while the **rastrigin** is a more difficult multi-modal problem given that the number of local minima grows exponentially with the increase of dimensionality (D). The differences between **sphere** and **rastrigin** are clearly shown in the two graphs at the bottom of Fig. 1.5.

Chapter 2
R Basics

2.1 Introduction

As explained in the preface of this book, the goal of this chapter is to briefly present the most relevant R tool aspects that need to be learned by non-experts in order to understand the examples discussed in this book. For a more detailed introduction to the tool, please consult (Paradis, 2002; Zuur et al., 2009; Venables et al., 2013).

R is a language and a computational environment for statistical analysis that was created by R. Ihaka and R. Gentleman in 1991 and that was influenced by the S and Scheme languages (Ihaka and Gentleman, 1996). R uses a high-level language, based on objects, that is flexible and extensible (e.g., by the development of packages) and allows a natural integration of statistics, graphics, and programming. The R system offers an integrated suite with a large and coherent collection of tools for data manipulation, analysis, and graphical display.

The tool is freely distributed under a GNU general public license and can be easily installed from the official web page (http://www.r-project.org), with several binary versions available for most commonly adopted operating systems (e.g., *Windows*, *macOS*). In particular, this book adopts the R 4.0.0 version that was launched in 2020, although the provided code should run in other versions. R can be run under the console (e.g., common in *Linux* systems) or using graphical user interface (GUI) applications (e.g., R for macOS). There are also several independent integrated development environments (IDE) for R, such as RStudio (https://rstudio.com/) and Jupyter notebook (https://jupyter.org/install). Figure 2.1 shows an example of the R console and GUI applications for the *macOS* system, and Fig. 2.2 exemplifies the RStudio and Jupyter IDEs.

R works mostly under a console interface (Fig. 2.1), where commands are typed after the prompt (`> `). An extensive help system is included. There are two console alternatives to get help on a particular function. For instance, `help(barplot)` or `?barplot` returns the help for the `barplot` function. It is also possible to search for a text expression, such as `help.search("linear models")`

P. Cortez, *Modern Optimization with R*, Use R!,
https://doi.org/10.1007/978-3-030-72819-9_2

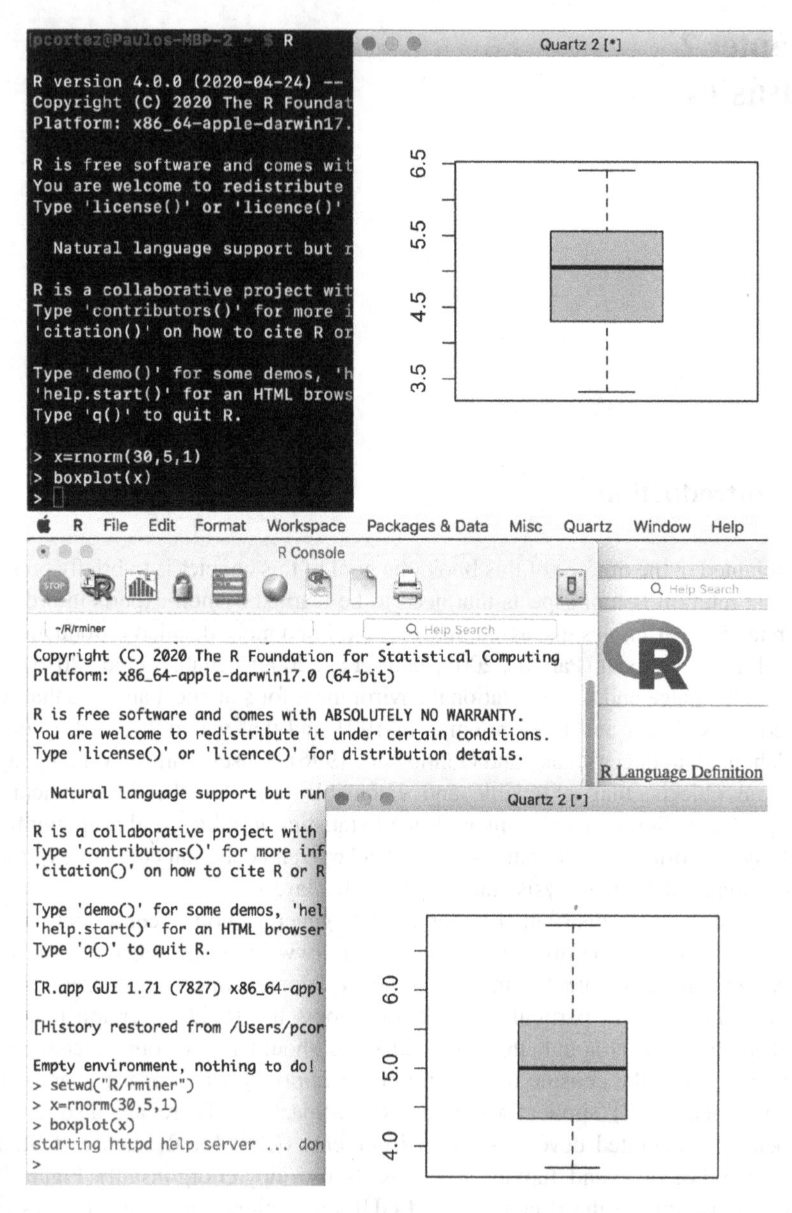

Fig. 2.1 Example of the R console (top) and GUI 4.0.0 versions (bottom) for *macOS*

or **??"linear models"**. For each function, the help system often includes a short description, the function main arguments, details, return value, references, and examples. The last item is quite useful for an immediate user perception of the function capabilities and can be accessed directly in the console, such

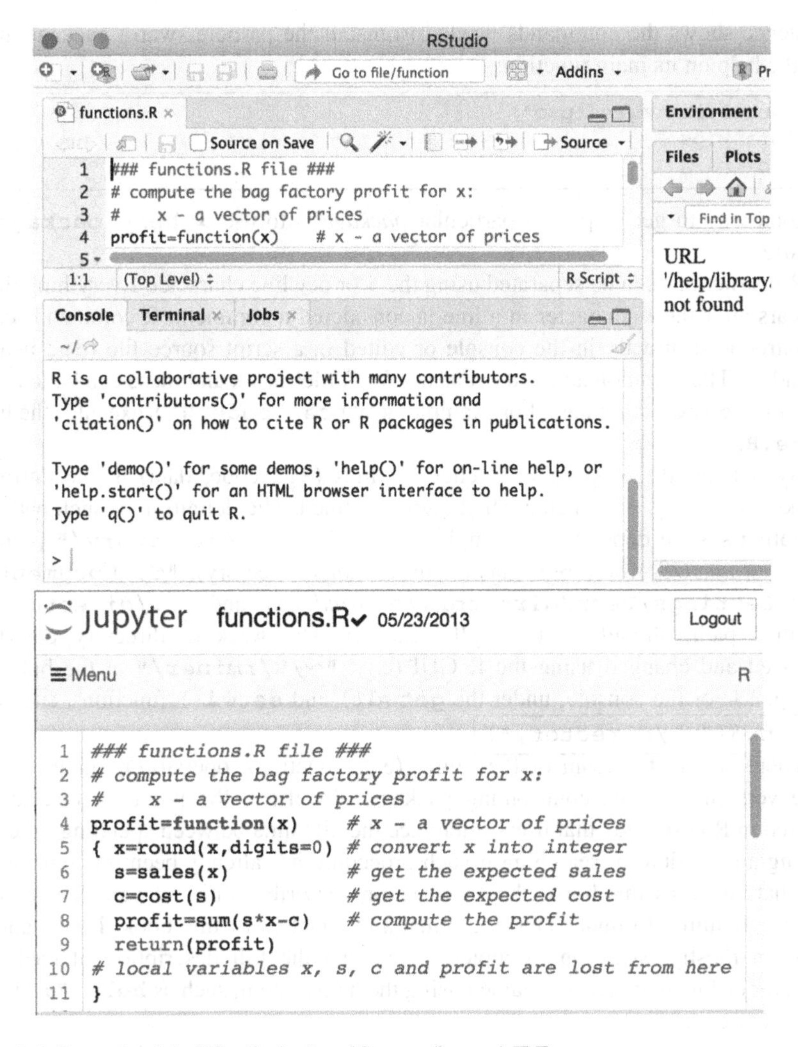

Fig. 2.2 Example of the RStudio (top) and Jupyter (bottom) IDEs

as by using `> example(barplot)`. Some demonstrations of interesting R scripts are available with the command **demo**, such as `> demo(graphics)` or `> demo()`.

The tool capabilities can be extended by installing packages. The full list of packages is available at the Comprehensive R Archive Network (CRAN): http://cran.r-project.org. Packages can be installed on the console, using the command **install.packages**, or GUI system, using the application menus. After installing a package, the respective functions and help are only available if the package is loaded using the **library** command. For example, the following sequence shows the commands needed to install the particle swarm package and get the help on its main function:

```
> install.packages("pso")
> library(pso)
> ?pso
```

A good way to get help on a particular *package* is to use `> help(package=`*package*`)`.

R instructions can be separated using the `;` or newline character. Everything that appears after the `#` character in a line is considered a comment. R commands can be introduced directly in the console or edited in a script source file (e.g., using RStudio). The common adopted extension for R files is `.R` and these can be loaded with the `source` command. For example, `source("code.R")` executes the file `code.R`.

By default, the R system searches for files (e.g., code, data) in the current working directory, unless an explicit path is defined. The definition of such path is operating system dependent. Examples of paths are: `"~/directory/"` (*Unix*, *Linux*, or *macOS*, where ~ means the user's home directory); `"C:/Documents and Settings/User/directory/"` (*Windows*); and `"../directory"` (relative path, should work in all systems). The working directory can be accessed and changed using the R GUI (e.g., `"~/R/rminer/"` at the bottom of Fig. 2.1) or the console, under the `getwd()` and `setwd()` functions, such as `> setwd("../directory")`.

There is a vast amount of R features (e.g., functions, operators), either in the base version or in its contributing packages. In effect, the number of features offered in R is so wide that users often face the dilemma between spending time in coding a procedure or searching if such procedure has already been implemented. In what concerns this book, the next sections describe some relevant R features that are required to understand the remaining chapters of this book. Explanation is given mostly based on examples, given that the full description of each R operator or function can be obtained using the help system, such as `help(":")` or `help("sort")`.

2.2 Basic Objects and Functions

R uses objects to store items, such as data and functions. The `=` (or `<-`[1]) operator can be used to assign an object to a variable. The class of an object is automatically assumed, with atomic objects including the `logical` (i.e., `FALSE`, `TRUE`), `character` (e.g., `"day"`), `integer` (e.g., `1L`), and `numeric` (e.g., `0.2`, `1.2e-3`) types. The type of any R object can be accessed by using the function `class`. There are also several constants defined, such as: `pi` – π; `Inf`

[1] Although the `<-` operator is commonly used in R, this book adopts the smaller `=` character.

– infinity; **NaN** – not a number; **NA** – missing value; and **NULL** – empty or null object.

The R system includes an extensive list of functions and operators that can be applied over a wide range of object types, such as:

- **class()** – gets type of object;
- **summary()** – shows a summary of the object;
- **print()** – shows the object;
- **plot()** – plots the object;
- **is.na()**, **is.nan()**, **is.null()** – check if object is **NA**, **NaN**, or **NULL**.

Another useful function is **ls()**, which lists all objects defined by the user. An example of a simple R session is shown here:

```
> s="hello world"
> print(class(s))    # character
[1] "character"
> print(s)           # "hello world"
[1] "hello world"
> x=1.1
> print(class(x))    # numeric
[1] "numeric"
> print(summary(x)) # summary of x
   Min. 1st Qu.  Median    Mean 3rd Qu.    Max.
    1.1     1.1     1.1     1.1     1.1     1.1
> plot(x)
> print(x)           # 1.1
[1] 1.1
> print(pi)          # 3.141593
[1] 3.141593
> print(sqrt(-1))    # NaN
[1] NaN
Warning message:
In sqrt(-1) : NaNs produced
> print(1/0)         # Inf
[1] Inf
```

There are also several containers, such as: **vector**, **factor**, **ordered**, **matrix**, **array**, **data.frame**, and **list**. Vectors, matrices, and arrays use an indexed notation (**[]**) to store and access several atoms. A factor is a special vector that contains only discrete values from a domain set, while ordered is a special factor whose domain levels have an order. A data frame is a special matrix where the columns (made of vectors or factors) have names. Finally, a list is a collection of distinct objects (called components). A list can include indexed elements (of any type, including containers) under the **[[]]** notation.

There is a large number of functions and operators that can be used to manipulate R objects (including containers). Some useful functions are:

- **c()** – concatenates several elements;
- **seq()** – creates a regular sequence;
- **sample()**, **runif()**, **rnorm()** – create random samples;

- **set.seed()** – sets the random generation seed number;
- **str()** – shows the internal structure of the object;
- **length(), sum(), mean(), median(), min(), max()** – compute the length, sum, average, median, minimum, or maximum of all elements of the object;
- **names()** – gets and sets the names of an object;
- **sort()** – sorts a vector;
- **which()** – returns the indexes of an object that follows a logical condition;
- **which.min(), which.max()** – return the index of the minimum or maximum value;
- **sqrt()** – square root of a number;
- **sin(), cos(), tan()** – trigonometric functions.

Examples of operators are:

- **$** – gets and sets a list component;
- **:** – generates regular sequences;
- **+, -, *, /** – simple arithmetic operators;
- **^** (or ******) – power operator;
- **%%** – rest of an integer division.

R also offers vast graphical based features. Examples of useful related functions are:

- **plot** – generic plotting function;
- **barplot** – bar plots;
- **pie** – pie charts;
- **hist** – histograms;
- **boxplot** – box-and-whisker plot;
- **wireframe** – 3D scatter plot (package **lattice**).

A graph can be sent to screen (default) or redirected to a device (e.g., PDF file). The description of all these graphical features is out of scope of this book, but a very interesting sample of R based graphs and code can be found at the R Graph Gallery https://www.r-graph-gallery.com.

An example of an R session that uses factors and vectors is presented next (Fig. 2.3 shows the graphs created by such code):

```
> f=factor(c("a","a","b","b","c")); print(f) # create factor
[1] a a b b c
Levels: a b c
> f[1]="c"; print(f)                           # change factor
[1] c a b b c
Levels: a b c
> print(levels(f))          # show domain levels: "a" "b" "c"
[1] "a" "b" "c"
> print(summary(f))                      # show a summary of y
a b c
1 2 2
> plot(f)                                     # show y barplot
```

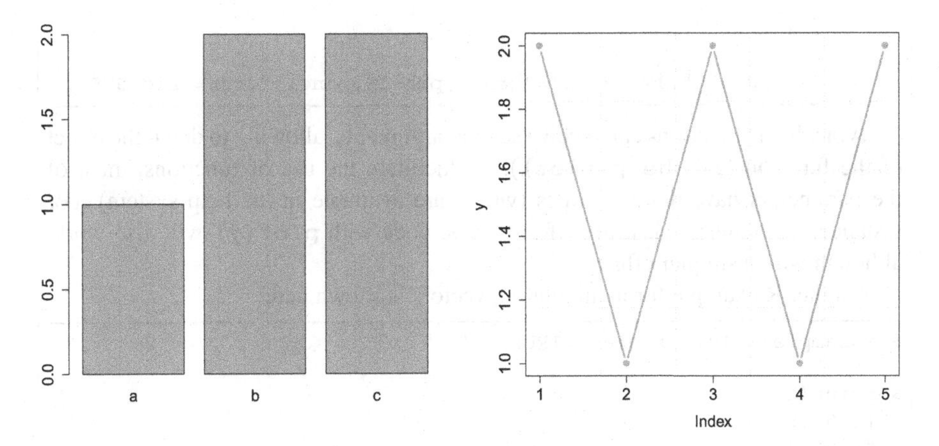

Fig. 2.3 Examples of a plot of a factor (left) and a vector (right) in R

```
> x=c(1.1,2.3,-1,4,2e-2) # creates vector x
> summary(x)             # show summary of x
   Min. 1st Qu.  Median    Mean 3rd Qu.    Max.
 -1.000   0.020   1.100   1.284   2.300   4.000
> print(x)               # show x
[1]  1.10  2.30 -1.00  4.00  0.02
> str(x)                 # show x structure
 num [1:5] 1.1 2.3 -1 4 0.02
> length(x)              # show the length of x
[1] 5
> x[2]                   # show second element of x
[1] 2.
> x[2:3]=(2:3)*1.1       # change 2nd and 3rd elements
> x[length(x)]=5         # change last element to 5
> print(x)               # show x
[1] 1.1 2.2 3.3 4.0 5.0
> print(x>3)             # show which x elements > 3
[1] FALSE FALSE  TRUE  TRUE  TRUE
> print(which(x>3))      # show indexes of x>3 condition
[1] 3 4 5
> names(x)=c("1st","2nd","3rd","4th","5th") # change names of x
> print(x)               # show x
1st 2nd 3rd 4th 5th
1.1 2.2 3.3 4.0 5.0
> print(mean(x))         # show the average of x
[1] 3.12
> print(summary(x))      # show a summary of x
   Min. 1st Qu.  Median    Mean 3rd Qu.    Max.
   1.10    2.20    3.30    3.12    4.00    5.00
> y=vector(length=5); print(y)      # FALSE, FALSE, ..., FALSE
[1] FALSE FALSE FALSE FALSE FALSE
> y[]=1; print(y)                   # all elements set to 1
[1] 1 1 1 1 1
> y[c(1,3,5)]=2; print(y)           # 2,1,2,1,2
```

```
[1] 2 1 2 1 2
> # fancier plot of y:
> plot(y,type="b",lwd=3,col="gray",pch=19,panel.first=grid(5,5))
```

Typically, R functions can receive several arguments, allowing to detail the effect of the function (e.g., **help(plot)**). To facilitate the use of functions, most of the parameters have default values (which are available in the help system). For instance, replacing the last line of the above code with **plot(y)** will also work, although with a simpler effect.

Another R example for manipulating vectors is shown here:

```
> x=sample(1:10,5,replace=TRUE)  # 5 random samples from 1 to 10
                                 #   with replacement
> print(x)                       # show x
[1] 10  5  5  1  2
> print(min(x))                  # show min of x
[1] 1
> print(which.min(x))      # show index of x that contains min
[1] 4
> print(sort(x,decreasing=TRUE)) # show x in decreasing order
[1] 10  5  5  2  1
> y=seq(0,20,by=2); print(y)     # y = 0, 2, ..., 20
 [1]  0  2  4  6  8 10 12 14 16 18 20
> print(y[x])                    # show y[x]
[1] 18  8  8  0  2
> print(y[-x])              # - means indexes excluded from y
[1]  4  6 10 12 14 16 20
> x=runif(5,0.0,10.0);print(x) # 5 uniform samples from 0 to 10
[1] 1.011359 1.454996 6.430331 9.395036 6.192061
> y=rnorm(5,10.0,1.0);print(y) # normal samples (mean 10, std 1)
[1] 10.601637  9.231792  9.548483  9.883687  9.591727
> t.test(x,y)                  # t-student paired test

   Welch Two Sample t-test

data:  x and y
t = -3.015, df = 4.168, p-value = 0.03733
alternative hypothesis: true difference in means is not equal to
   0
95 percent confidence interval:
 -9.2932638 -0.4561531
sample estimates:
mean of x mean of y
 4.896757  9.771465
```

The last R function (**t.test()**) checks if the differences between the x and y averages are statistically significant. Other statistical tests are easily available in R, such as **wilcox.test** (Wilcoxon) or **chisq.test** (Pearson's chi-squared). In the above example, x is created using a uniform distribution $\mathcal{U}(0, 10)$, while y is created using a normal one, i.e., $\mathcal{N}(10, 1)$. Given that R is a strong statistical tool, there is an extensive list of distribution functions (e.g., binomial, Poisson), which can be accessed using **help("Distributions")**.

The next R session is about **matrix** and **data.frame** objects:

```
> m=matrix(ncol=3,nrow=2); m[,]=0; print(m) # 3x2 matrix
     [,1] [,2] [,3]
[1,]    0    0    0
[2,]    0    0    0
> m[1,]=1:3; print(m)                        # change 1st row
     [,1] [,2] [,3]
[1,]    1    2    3
[2,]    0    0    0
> m[,3]=1:2; print(m)                        # change 3rd column
     [,1] [,2] [,3]
[1,]    1    2    1
[2,]    0    0    2
> m[2,1]=3; print(m)                         # change m[2,1]
     [,1] [,2] [,3]
[1,]    1    2    1
[2,]    3    0    2
> print(nrow(m))                             # number of rows
[1] 2
> print(ncol(m))                             # number of columns
[1] 3
> m[nrow(m),ncol(m)]=5; print(m)             # change last element
     [,1] [,2] [,3]
[1,]    1    2    1
[2,]    3    0    5
> m[nrow(m)-1,ncol(m)-1]=4; print(m)         # change m[1,2]
     [,1] [,2] [,3]
[1,]    1    4    1
[2,]    3    0    5
> print(max(m))                              # show maximum of m
[1] 5
> m=sqrt(m); print(m)                        # change m
         [,1] [,2]     [,3]
[1,] 1.000000    2 1.000000
[2,] 1.732051    0 2.236068
> m[1,]=c(1,1,2013); m[2,]=c(2,2,2013)       # change m
> d=data.frame(m)                            # create data.frame
> names(d)=c("day","month","year")           # change names
> d[1,]=c(2,1,2013); print(d)                # change 1st row
  day month year
1   2     1 2013
2   2     2 2013
> d$day[2]=3; print(d)                       # change d[1,2]
  day month year
1   2     1 2013
2   3     2 2013
> d=rbind(d,c(4,3,2014)); print(d)           # add row to d
  day month year
1   2     1 2013
2   3     2 2013
3   4     3 2014
> # change 2nd column of d to factor, same as d[,2]=factor(...
> d$month=factor(c("Jan","Feb","Mar"))
```

```
> print(summary(d))                          # summary of d
      day           month          year     
 Min.   :2.0    Feb:1   Min.   :2013  
 1st Qu.:2.5    Jan:1   1st Qu.:2013  
 Median :3.0    Mar:1   Median :2013  
 Mean   :3.0            Mean   :2013  
 3rd Qu.:3.5            3rd Qu.:2014  
 Max.   :4.0            Max.   :2014  
```

The last section example is related with lists:

```
> l=list(a="hello",b=1:3) # list with 2 components
> print(summary(l))       # summary of l
  Length Class  Mode     
a 1      -none- character
b 3      -none- numeric  
> print(l)                # show l
$a
[1] "hello"

$b
[1] 1 2 3

> l$b=l$b^2+1;print(l)      # change b to (b*b)+1
$a
[1] "hello"

$b
[1]  2  5 10

> v=vector("list",3)  # vector list
> v[[1]]=1:3          # change 1st element of v
> v[[2]]=2            # change 2nd element of v
> v[[3]]=l            # change 3rd element of v
> print(v)            # show v
[[1]]
[1] 1 2 3

[[2]]
[1] 2

[[3]]
[[3]]$a
[1] "hello"

[[3]]$b
[1] 2 3 4

> print(length(v))    # length of v
[1] 3
```

2.3 Controlling Execution and Writing Functions

The R language contains a set of control-flow constructs that are quite similar to other imperative languages (e.g., C, Java). Such constructs can be accessed using the console command line **help("Control")** and include:

- **if** (*condition*) *expression* – if *condition* is true then execute *expression*;
- **if** (*condition*) *expression1* **else** *expression2* – another conditional execution variant, where *expression1* is executed if *condition* is **TRUE**, else *expression2* is executed;
- **switch(...)** – conditional control function that evaluates the first argument and based on such argument chooses one of the remaining arguments;
- **for** (*variable* **in** *sequence* **)** *expression* – cycle where *variable* assumes in each iteration a different value from *sequence*;
- **while** (*condition*) *expression* – loop that is executed while *condition* is true;
- **repeat** *expression* – executes expression (stops only if there is a break);
- **break** – breaks out a loop;
- **next** – skips to next iteration.

A *condition* in R is of the **logical** type, assumed as the first **TRUE** or **FALSE** value. Similarly to other imperative languages, several logical operators can be used within a condition (use **help("Logic")** for more details):

- *x*==*y*, *x*!=*y*, *x*>*y*, *x*>=*y* *x*<*y*, and *x*>=*y* – equal, different, higher, higher or equal, lower, lower or equal to;
- !*x* – negation of *x*;
- *x*&*y* and *x*|*y* – *x* and *y* element-wise logical AND and OR (may generate several **TRUE** or **FALSE** values);
- *x*&&*y* and *x*||*y* – left to right examination of logical AND and OR (generates only one **TRUE** or **FALSE** value);
- **xor** (*x*, *y*) – element-wise exclusive OR.

Regarding the *expression*, it can include a single command, $expression_1$, or a compound expression, under the form { $expression_1$; ... $expression_n$ }. An R session is presented here to exemplify how control execution is performed:

```
# two if else examples:
> x=0; if(x>0) cat("positive\n") else if(x==0) cat("neutral\n")
    else cat("negative\n")
neutral
> if(xor(x,1)) cat("XOR TRUE\n") else cat("XOR FALSE\n")
XOR TRUE
> print(switch(3,"a","b","c"))          # numeric switch example
[1] "c"
> x=1; while(x<3) { print(x); x=x+1;} # while example
[1] 1
[1] 2
> for(i in 1:3) print(2*i)             # for example #1
[1] 2
```

```
[1] 4
[1] 6
> for(i in c("a","b","c")) print(i)    # for example #2
[1] "a"
[1] "b"
[1] "c"
> for(i in 1:10) if(i%%3==0) print(i) # for example #3
[1] 3
[1] 6
[1] 9
                                    # character switch example:
> var="sin";x=1:3;y=switch(var,cos=cos(x),sin=sin(x))
> cat("the",var,"of",x,"is",round(y,digits=3),"\n")
the sin of 1 2 3 is 0.841 0.909 0.141
```

This example introduces two new functions: **cat** and **round**. Similarly to **print**, the **cat** function concatenates and outputs several objects, where **"\n"** means the newline character,[2] while **round** rounds the object values with the number of digits defined by the second argument.

The element-wise logical operators are useful for filtering containers, such as shown in this example:

```
> x=1:10;print(x)
 [1]  1  2  3  4  5  6  7  8  9 10
> print(x>=3&x<8)                          # select some elements
 [1] FALSE FALSE  TRUE  TRUE  TRUE  TRUE  TRUE FALSE FALSE FALSE
> I=which(x>=3&x<8);print(I)               # indexes of selection
[1] 3 4 5 6 7
> d=data.frame(x=1:4,f=factor(c(rep("a",2),rep("b",2))))
> print(d)
  x f
1 1 a
2 2 a
3 3 b
4 4 b
> print(d[d$x<2|d$f=="b",])                # select rows
  x f
1 1 a
3 3 b
4 4 b
```

The **rep** function replicates elements of vectors and lists. For instance, **rep(1:3,2)** results in the vector: **1 2 3 1 2 3**.

The power of R is highly enhanced by the definition of functions, which similarly to other imperative languages (e.g., C, Java) define a portion of code that can be called several times during program execution. A function is defined as: *name*=**function(***arg1*, *arg2*, **...)** *expression*, where: *name* is the function name; *arg1*, *arg2*, **...** are the arguments; and *expression* is a single command

[2]For C language users, there is also the sprintf function (e.g., **sprintf("float: %.2f string: %s",pi,"pi")**).

or compound expression. Arguments can have default values by using *arg=arg expression* in the function definition. Here, *arg expression* can be a constant or an arbitrary R expression, which can use other arguments from the same function. The three dots special argument (`...`) means several arguments that are passed on to other functions (an example of `...` usage is shown in Sect. 3.2).

The scope of a function code is local, meaning that any assignment (`=`) within the function is lost when the function ends. If needed, global assignment in R is possible using the `<<-` operator (see Sect. 4.5 for an example). Also, a function can only return one object, which is defined by the value set under the **`return`** command or, if not used, last line of the function. Functions can be recursive, i.e., a function that calls the same function but typically with different arguments. Moreover, a function may define other functions within itself.

As an example, the following code was edited in a file named **`functions.R`** and computes the profit for the **bag prices** task (Sect. 1.8) and defines a recursive function:

```
### functions.R file ###
# compute the bag factory profit for x:
#    x - a vector of prices
profit=function(x)      # x - a vector of prices
{ x=round(x,digits=0) # convert x into integer
  s=sales(x)           # get the expected sales
  c=cost(s)            # get the expected cost
  profit=sum(s*x-c)    # compute the profit
  return(profit)
# local variables x, s, c and profit are lost from here
}

# compute the cost for producing units:
#    units - number of units produced
#    A - fixed cost, cpu - cost per unit
cost=function(units,A=100,cpu=35-5*(1:length(units)))
{ return(A+cpu*units) }

# compute the estimated sales for x:
#    x - a vector of prices, m - marketing effort
#    A, B, C - constants of the estimated function
sales=function(x,A=1000,B=200,C=141,
               m=seq(2,length.out=length(x),by=-0.25))
{ return(round(m*(A/log(x+B)-C),digits=0))}

# example of a simple recursive function:
fact=function(x=0) # x - integer number
{ if(x==0) return(1) else return(x*fact(x-1))}
```

In this example, although object **`x`** is changed inside function **`profit`**, such change is not visible outside the function scope. The code also presents several examples of default arguments, such as constants (e.g., **`C=141`**) and more complex expressions (e.g., **`cpu=35-5*(1:length(units))`**). The last function (**`fact`**) was included only for demonstrative purposes of a recursive function, since it only works for single numbers. It should be noted that R includes the enhanced **factorial** function that works with both single and container objects.

When invoking a function call, arguments can be given in any order, provided the argument name is explicitly used, under the form *argname=object*, where *argname* is the name of the argument. Else, arguments are assumed from left to right. The following session loads the previous code and executes some of its functions:

```
> source("functions.R") # load the code
> cat("class of profit is:",class(profit),"\n") # function
class of profit is: function
> x=c(414.1,404.2,408.3,413.2,395.0)
> y=profit(x); cat("maximum profit:",y,"\n")
maximum profit: 43899
> cat("x is not changed:",x,"\n")
x is not changed: 414.1 404.2 408.3 413.2 395
> cat("cost(x=",x,")=",cost(x),"\n")
cost(x= 414.1 404.2 408.3 413.2 395 )= 12523 10205 8266 6298 4050
> cat("sales(x=",x,")=",sales(round(x)),"\n")
sales(x= 414.1 404.2 408.3 413.2 395 )= 30 27 23 19 16
> x=c(414,404) # sales for 2 bags:
> cat("sales(x=",x,")=",sales(x),"\n")
sales(x= 414 404 )= 30 27
> cat("sales(x,A=1000,m=c(2,1.75))=",sales(x,1000,m=c(2,1.75)),"\
   n")
sales(x,A=1000,m=c(2,1.75))= 30 27
> # show 3! :
> x=3; cat("fact(",x,")=",fact(x),"\n")
fact( 3 )= 6
```

R users tend to avoid the definition of loops (e.g., **for**) in order to reduce the number of lines of code and mistakes. Often, this can be achieved by using special functions that execute an argument function over all elements of a container (e.g., vector or matrix), such as: **sapply** or **lapply**, which runs over a vector or list; and **apply**, which runs over a matrix or array. An example that demonstrates these functions is shown next:

```
> source("functions.R")           # load the code
> x=1:5                           # show the factorial of 1:5
> cat(sapply(x,fact),"\n")        # fact is a function
1 2 6 24 120
> m=matrix(ncol=5,nrow=2)
> m[1,]=c(1,1,1,1,1)              # very cheap bags
> m[2,]=c(414,404,408,413,395)    # optimum
# show profit for both price setups:
> y=apply(m,1,profit); print(y) # profit is a function
[1] -7854 43899
```

The second argument of **apply()** is called **MARGIN** and indicates if the function (third argument) is applied over the rows (**1**), columns (**2**), or both (**c(1,2)**).

2.4 Importing and Exporting Data

The R tool includes several functions for importing and exporting data. Any R object can be saved into a binary or ASCII (using an R external representation) with the **save** function and then loaded with the **load** command. All R session objects, i.e., the current workspace, can be saved with **save.image()**, which also occurs with **q("yes")** (quitting function). Such workspace is automatically saved into a **.RData** file. Similarly to reading R source files, file names are assumed to be found in the current working directory (corresponding to **getwd()**), unless the absolute path is specified in the file names.

Text files can be read by using the **readLines** (all files or one line at the time) function. A text file can be generated by using a combination of: **file** (create or open a connection), **writeLines** (write lines into a connection) and **close** (close a connection); or **sink** (divert output to a connection) and console writing functions (e.g., **print** or **cat**).

The next example shows how to save/load objects and text files:

```
> x=list(a=1:3,b="hello!")              # x is a list
> save(x,file="x.Rdata",ascii=TRUE) # save into working directory
> rm(x)                                 # remove an object
> print(x)                              # gives an error
Error in print(x) : object 'x' not found
> load("x.Rdata")                       # x now exists!
> print(x)                              # show x
$a
[1] 1 2 3

$b
[1] "hello!"

> t=readLines("x.Rdata")                # read all text file
> cat("first line:",t[1],"\n")          # show 1st line
first line: RDA3
> cat("first line:",readLines("x.Rdata",n=1),"\n")
first line: RDA3
> # write a text file using writeLines:
> conn=file("demo.txt")                # create a connection
> writeLines("hello!", conn)            # write something
> close(conn)                           # close connection
> # write a text file using sink:
> sink("demo2.txt")                     # divert output
> cat("hello!\n")                       # write something
> sink()                                # stop sink
```

A common way of loading data is to read tabular or spreadsheet data (e.g., CSV format) by using the **read.table** function (and its variants, such as **read.csv**). The reverse operation is performed using the **write.table** command. A demonstrative example for reading and writing tabular data is shown here:

```
> # create and write a simple data.frame:
> d=data.frame(day=1:2,mon=factor(c("Jan","Feb")),year=c(12,13))
> print(d)
  day mon year
1   1 Jan   12
2   2 Feb   13
> write.table(d,file="demo.csv",row.names=FALSE,sep=";")
> # read the created data.frame:
> d2=read.table("demo.csv",header=TRUE,sep=";")
> print(d2)
  day mon year
1   1 Jan   12
2   2 Feb   13
> # read white wine quality dataset from UCI repository:
> library(RCurl)
> URL="http://archive.ics.uci.edu/ml/machine-learning-databases/
    wine-quality/winequality-white.csv"
> wine=getURL(URL)
# write "winequality-white.csv" to working directory:
> write(wine,file="winequality-white.csv")
# read file:
> w=read.table("winequality-white.csv",header=TRUE,sep=";")
> cat("wine data (",nrow(w),"x",ncol(w),")\n") # show nrow x ncol
wine data ( 4898 x 12 )
```

The "R data import/export" section of the R manual, accessible via **help.start()**, includes a wide range of data formats that can be imported by installing the **foreign** package, such as **read.spss** (SPSS format), **read.mtp** (Minitab Portable Worksheet format), **read.xport** (SAS XPORT format), **read.arff** (ARFF Weka data mining tool format), and **read.octave** (Octave text data format). Other file formats can be read using other packages, such as:

- Excel files: using the **readxl** package and function **read_excel** or using **gdata** package and function **read.xls**;
- Web content: using **RCurl** package and **getURL** function;
- JSON objects: using **rjson** package and **fromJSON** and **toJSON** functions;
- databases: MySQL using package **RMySQL**;
- text corpus: using the **tm** package.

The Rstudio data import cheat sheet condenses some of these and other ways to import data into R: https://rawgit.com/rstudio/cheatsheets/master/data-import.pdf.

Any R graphic can be saved into a file by changing the output device driver, creating the graphic and then closing the device, via **dev.off()**. Several graphic devices are available, such as **pdf**, **png**, **jpeg**, and **tiff**. The next example shows the full code used to create the top left graph of Fig. 1.5:

```
# create PDF file:
DIR=""                  # change if different directory is used
pdf(paste(DIR,"sumbits.pdf",sep=""),width=5,height=5)

sumbinint=function(x)          # sum of bits of an integer
```

```
{ return(sum(as.numeric(intToBits(x))))}

sumbits=function(x)                # sum of bits of a vector
{ return(sapply(x,sumbinint))}

D=8; x=0:(2^D-1) # x is the search space (integer representation)
y=sumbits(x)     # y is the number of binary bits of x
plot(x,y,type="l",ylab="evaluation function",
     xlab="search space (x)",lwd=2)
pmax=c(x[which.max(y)],max(y)) # maximum point coordinates
points(pmax[1],pmax[2],pch=19,lwd=2)  # plot maximum point
legend("topleft","optimum",pch=19,lwd=2)    # add a legend
dev.off()                              # close the device
```

This example introduces the functions: **intToBits**, which converts an integer into a binary representation; **as.numeric**, which converts an object into numeric; and **legend**, which adds legends to plots.

2.5 Additional Features

This section discusses several additional R features: command line execution of R, parallel computing, getting R source code of a function, interfacing with other computer languages, automatic generation of documentation, and interactive web apps.

Command Line Execution

The R environment can be executed directly from the operating system console, under two possibilities:

- **R [options] [< infile] [> outfile]; or**
- **R CMD command [arguments].**

The full details can be accessed by running `$ R --help` in the operating system console (**$** is the *macOS* prompt). This direct mode can be used for several operations, such as compiling files for use in R or executing an R file in batch processing, without manual intervention. For example, if the previous shown code for creating a pdf image file is saved in a file called **ex-sumbits.R**, then such code can be directly executed in R by using: `$ R --vanilla --slave < ex-sumbits.R`.

Parallel Computing

High-performance and parallel computing is an advanced computer science topic. In this book, only simpler concepts are introduced. Parallelization assumes that a computer task can use a particular computational infrastructure to process data in parallel. The goal is to improve the computing performance (e.g., total time elapsed) by splitting a computer program into smaller pieces (called threads) that can be independently solved using different computing processors. Modern computers (e.g., desktop or laptops) have multiple cores (n processors) that can be used to simultaneously process different tasks. For more intensive computation there are also computer clusters, which connect several computing machines that work together as if they were a single unit with n processors.

Some tasks are easy to parallelize. For instance, any cycle execution (e.g., **for**, **while**, **lapply**) with n iterations can be executed using n different processors (or cores), provided there are no dependencies among the different iterations. This is applicable to a wide range of population based optimization methods that require the independent evaluation of several solutions, such as generic algorithms (Sect. 5.9). Moreover, due to system core processing and data memory communication costs, the speedup (improvement in execution time) can be less than what would be linearly expected when increasing the number of processors. For instance, if task A requires 10 s, then the parallel version of A can require more than 1 s when using $n = 10$ processors.

There is a CRAN task view for high-performance and parallel computing with R (http://cran.r-project.org/web/views/HighPerformanceComputing.html). The view includes several packages that are useful for high-performance computing. In particular, the **parallel** package is widely used in R and it provides multiple useful functions. The **par1.R** R file example uses the **parLapply** function, which implements a parallel variant of the **lapply** function that cycles a container:

```
### par1.R file ###
# parallel example with "parLapply" function:

# load the libraries:
library(parallel)
library(tictoc)

cl=makeCluster(getOption("cl.cores", 2)) # set 2 cores
mysleep=function(s) { Sys.sleep(s) }      # my sleep function
x=c(2,2) # in seconds

tic("single execution:") # set the timer
lapply(x,mysleep) # execute serial cycle
toc() # measure time elapsed

tic("parallel execution:") # set the timer
parLapply(cl,x,mysleep) # execute parallel cycle
toc() # measure time elapsed
```

```
stopCluster(cl) # stop the cluster
```

The code also loads the **tictoc** package that is useful to measure time elapsed executions in R via two simple functions: **tic** (start a timer) and **toc** (compute time elapsed since last **tic** call). The code first creates a **cluster** object that assumes 2 processing cores via the **makeCluster** function. Then, it defines a sleeping function that puts the processor idle during **s** seconds via the **Sys.sleep** R function. Next, a simple serial execution is performed, via **lapply** function, followed by the parallelized version executed using **parLapply()**. Finally, the cluster is stopped using the **stopCluster** function. The obtained result in a *macOS* laptop is:

```
> source("par1.R")
single execution:: 4.008 sec elapsed
parallel execution:: 2.033 sec elapsed
```

The example exhibits a linear speedup, where the parallel version requires just half of the serial execution time.

Often, some pieces of code require more execution time than others, meaning that the problem is unbalanced. In such cases, the **parLapplyLB** can be used as a more efficient variant of the **parLapply** function, since it automatically performs a load balancing, aiming to achieve an automatic efficient parallel distribution of the tasks. The file **par2.R** exemplifies this:

```
### par2.R file ###
# parallel example with "parLapply" function:

# load the libraries:
library(parallel)
library(tictoc)

n=detectCores() # attempt to detect the CPU cores
cat("Set the cluster with",n,"cores.\n")
cl=makeCluster(n) # set the cores


V=vector("list",n)
set.seed(123) # set for replicability
s=sample(1:100000000,n) # variable sizes
for(i in 1:n) V[[i]]=1:s[i] # sequence

# serial execution:
tic("serial execution")
lapply(V,mad)
toc()

# parallel execution:
tic("parallel execution:")
parLapplyLB(cl,V,mad) # execute parallel cycle
toc()
```

```
# load balancing parallel execution:
tic("LB parallel execution:")
# load balancing parallel version:
parLapply(cl,V,mad)
toc()

stopCluster(cl) # stop the cluster
```

The code creates a vector list **V** with **n** elements (the number of cores, automatically detected using **detectCores**). The vector is filled with a variable sequence (1,...,**s[i]**) whose size depends on the value of the **s** sample. In this example, the **set.seed** R function was used to initialize the sample random generator, such that any computer that executes this code obtains the same **s** object. The code uses the **mad** R function, which computes the median absolute deviation of a numeric vector. The higher the vector size, the larger will be the computational effort when computing the **mad** value. Given that object **V** contains different sequence sizes, the **mad** effort for each sequence is also substantially different. The **par2.R** code first executes a serial computation, then a simple parallel one and finally the load balanced version. Using the same *macOS* laptop, the result is:

```
> source("par2.R")
Set the cluster with 8 cores.
serial execution: 28.971 sec elapsed
parallel execution:: 12.22 sec elapsed
LB parallel execution:: 10.043 sec elapsed
```

In this case, **n=8** cores were detected. While the simpler parallelization reduced the total time elapsed, the load balancing version obtained a lower execution time.

Source Code of a Function

Given that all functions are stored as objects, it is easy in R to access the full code of any given function, including built-in functions, by using the **methods** and **getAnywhere** commands, such as:

```
> methods(mean) # list all available methods for mean function
[1] mean.Date     mean.POSIXct  mean.POSIXlt  mean.default
    mean.difftime
see '?methods' for accessing help and source code
> getAnywhere(mean.default) # show R code for default mean
    function
A single object matching mean.default was found
It was found in the following places
  package:base
  registered S3 method for mean from namespace base
  namespace:base
with value
```

```
function (x, trim = 0, na.rm = FALSE, ...)
{
    if (!is.numeric(x) && !is.complex(x) && !is.logical(x)) {
        warning("argument is not numeric or logical: returning NA
            ")
        return(NA_real_)
    }
    if (na.rm)
        x <- x[!is.na(x)]
    if (!is.numeric(trim) || length(trim) != 1L)
        stop("'trim' must be numeric of length one")
    n <- length(x)
    if (trim > 0 && n) {
        if (is.complex(x))
            stop("trimmed means are not defined for complex data"
                )
        if (anyNA(x))
            return(NA_real_)
        if (trim >= 0.5)
            return(stats::median(x, na.rm = FALSE))
        lo <- floor(n * trim) + 1
        hi <- n + 1 - lo
        x <- sort.int(x, partial = unique(c(lo, hi)))[lo:hi]
    }
    .Internal(mean(x))
}
<bytecode: 0x7fd643648560>
<environment: namespace:base>
```

Interfacing with Other Languages

The R environment can interface with other programming languages, such as Fortran, C, Java, and Python. This book presents an interface example with the Python language. Further examples can be found in:

- C and Fortran – http://adv-r.had.co.nz/C-interface.html and the "Writing R Extensions" user manual, by typing **help.start()**;
- Java – http://www.rforge.net/rJava/.
- Python – https://rstudio.github.io/reticulate/.

The **reticulate** R package allows to interface with the Python language. Any Python module (logical organization of Python code, similar to an R package) can be imported via the **import()** function. And a Python script can be sourced using the **source_python()** function. The example executes the mean absolute error (MAE) metric (popular in regression tasks):

$$MAE = \frac{\sum_{i=1}^{n} |y_i - x_i|}{n} \tag{2.1}$$

where n is the length of the numeric vectors $y = (y_1, y_2, \ldots)$ and $x = (x_1, x_2, \ldots)$. There needs to be a Python installation in the computer system where the R code is executed, which can be checked by executing this code in the R console: **> Sys.which("python")**. If needed, the Python path can be changed using the **use_python()** command. The simple Python script is named **mae.py** and it uses the **numpy** (numerical computing) module:

```
### mae.py Python script file ###
import numpy as np # np is a shortcut for numpy

# definition of the simple MAE Python function:
def mae_py(y,x):
  y=np.array(y) # set y as a numeric array
  x=np.array(x) # set x as a numeric array
  mae=np.mean(abs(y-x)) # mae calculation
  return mae # return the mae value
```

The R file that calls the Python script is **mae.R**:

```
### mae.R file ###

library(reticulate) # load the package

# the simple MAE function:
mae_R=function(y,x) mean(abs(y-x))

source_python("mae.py")  # source the python script

a=c(0,1,0.5); b=c(1,1,1) # two numeric vectors
mae1=mae_py(a,b)         # compute mae using python
mae2=mae_R(a,b)          # compute mae using R

cat("the mae is",mae1,"=",mae2,"\n")
```

The R file creates two numeric vectors (**a** and **b**) and then computes the MAE using the Python **mae_py** and R **mae_R** functions. The result is:

```
> source("mae.R")
the mae is 0.5 = 0.5
```

As expected, both functions compute the same $MAE = 0.5$ value.

Interactive Web Applications

The **shiny** package allows an easy creation of interactive web applications. If needed, the package can be installed using **install.packages("shiny")**. The package can be run in the standard R console or RStudio IDE, with the latter environment providing additional functionalities. A shiny application is contained in a single file named **app.R** that is inserted into a directory or folder, such as

demoApp. The **app.R** file should contain three components: an user interface (UI) object, a server function, and a call to the **shinyApp** function. The R shiny tutorial web page https://shiny.rstudio.com/tutorial contains several shiny templates and examples. This section presents just a brief **shiny** demonstration (see Sect. 7.4 for another example), which assumes that the file **app.R** is inserted into the directory **demoApp**:

```
### demoApp/app.R file ###

library(shiny) # load the package

# User Interface (UI): with 2 inputs and 1 output
ui=fluidPage( # begin fluidPage
  titlePanel("demoApp"), # title of the panel
  sidebarLayout( # sidebar with input and outputs
  sidebarPanel(  # panel for inputs
    selectInput("var", # input: 3 character choices
      label="Choose a variable",
      choices=c("pH","alcohol","quality"),
      selected="pH" # default choice
    ),
    sliderInput( # input: numeric selection
      inputId="bins",
      label="number of bins:",
      min=2,    # minimum value
      max=30, # maximum value
      value=10 # default value
    )), # end sidebarPanel
  mainPanel( # panel for outputs
      plotOutput(outputId="distPlot")
    )) # end sidebar
) # end fluidPage

# Server function:
server=function(input,output)
{
  output$distPlot=renderPlot({
    x=w[,input$var] # select variable from w
    # draw the histogram with the specified number of bins
    text=paste("Histogram of",input$var) # histogram title
    hist(x,breaks=input$bins,col="gray",main=text)
  })
}

# load the wine data and create object w
file="https://archive.ics.uci.edu/ml/machine-learning-databases/
    wine-quality/winequality-white.csv"
w=read.table(file,sep=";",heade=TRUE)
# call to shinyApp
shinyApp(ui=ui,server=server)
```

The goal of the demonstration is to plot the histogram of three white wine quality variables, also allowing to control the number of bins that are visualized in the histogram. The UI is created using the **fluidPage** function and it contains several components:

- **titlePanel** – sets the title of the panel;
- **sidebarLayout** – defines a sidebar with input and output components;
- **sidebarPanel** – sets a panel with the input components;
- **selectInput** – selects a text input that defines the analyzed wine variable;
- **sliderInput** – selects a numeric input that sets the number of bins of the histogram;
- **mainPanel** – sets the main area with the output component;
- **plotOutput** – renders the output plot.

The respective user selections are automatically stored in the **input$var** (text) and **input$bins** (numeric value) objects. The server function controls the input to output execution. Each time the user changes an input, the function **renderPlot** is automatically activated. The function selects a variable from the **w** **data.frame**, defines the histogram title, and plots the respective histogram. The histogram text title is created by using the **paste** function, which merges several components (text or numbers) into a single character. Before calling the **shinyApp** function, the demonstration file reads the wine data into the **w** object, which is now accessible (as a global variable) inside the **server** function.

The interactive application is run in a browser when the command **runApp** is executed in the R console or RStudio IDE:

```
> library(shiny)
> runApp("demoApp")
```

Figure 2.4 shows two example executions of the interactive demonstration.

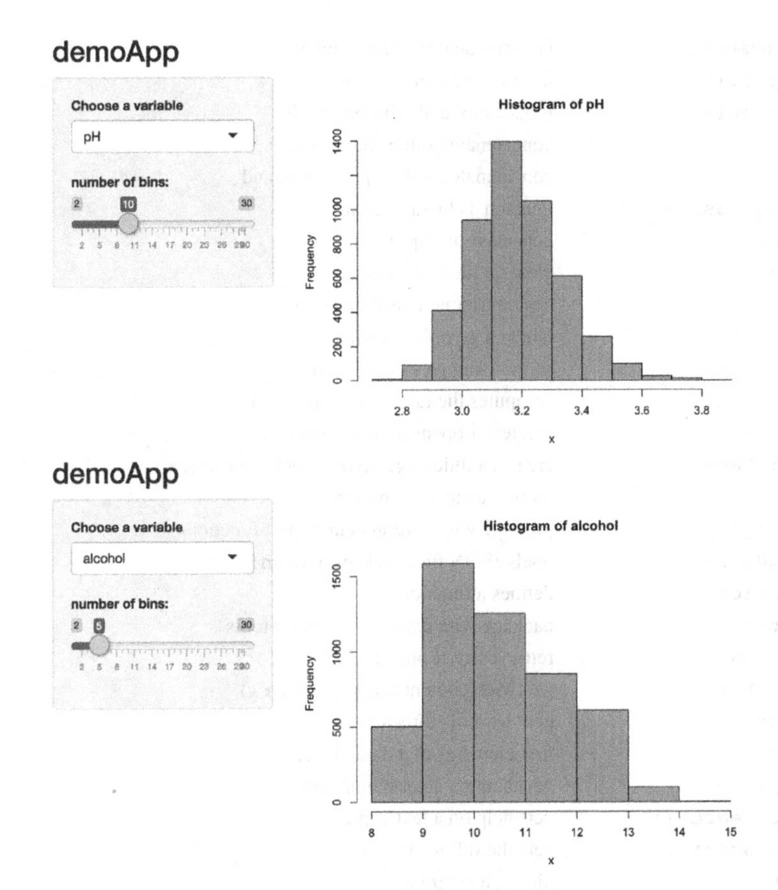

Fig. 2.4 Example of two executions of the wine quality interactive application

2.6 Command Summary

`Inf`	infinity value
`NA`	missing value
`NULL`	empty or null object
`NaN`	not a number constant
`RCurl`	package for network (HTTP/FTP/...) interface
`RMySQL`	package to interface with MySQL
`Sys.sleep()`	suspends execution
`Sys.which()`	returns full path names of an executable
`apply()`	applies a function over a matrix or array

as.numeric()	converts an object into numeric
barplot()	draws a bar plot
boxplot()	plots a box-and-whisker graph
c()	concatenates values into a vector
cat()	concatenates and outputs command
chisq.test()	Pearson's chi-squared test
class()	gets class of object
close()	closes a file connection
cos()	cosine trigonometric function
dev.off()	closes a graphical device
example()	shows examples of a command
factorial()	computes the factorial of an object
file()	creates or opens a file connection
fluidPage()	creates a fluid page layout (package **shiny**)
for()	loops execution command
foreign	package with several data import functions
fromJSON()	reads JSON file (package **rjson**)
function()	defines a function
gdata	package with data manipulation tools
getAnywhere()	retrieves an R object
getURL()	gets Web content (package **RCurl**)
getwd()	gets working directory
head()	first elements of a data object
help()	gets help on a particular subject
help.search()	gets help on a text expression
help.start()	gets the full R manual
hist()	plots a histogram
if()	conditional execution command
import()	imports a Python module for calling from R (package **reticulate**).
install.packages()	installs a package
intToBits()	converts integer into binary representation
is.na()	checks missing data
is.nan()	checks if NaN
is.null()	checks if NULL
jpeg()	sets graphical device to jpeg file
lapply()	applies a function over a list or vector
lattice	package with high-level data visualization functions
legend()	adds a legend to a plot
length()	number of elements of an object
library()	loads a package
load()	loads an object from file

ls()	lists created objects
mainPanel()	sets main area (package **shiny**)
makeCluster()	creates a parallel cluster (package **parallel**)
max()	maximum of all values
mean()	mean of all values
median()	median of all values
methods()	lists methods for functions
min()	minimum of all values
names()	gets and sets the names of an object
parLapply()	parallel computation (package **parallel**)
parLapplyLB()	parallel computation (package **parallel**)
parallel	package for parallel computation
paste()	concatenates elements into a character
pdf()	sets graphical device to pdf file
pie()	plots a pie chart
pi	mathematical π value
plot()	generic plot of an object
plotOutput()	renders a plot (package **shiny**)
png()	sets graphical device to png file
print()	shows an object
read.arff()	reads an ARFF Weka file (package **foreign**)
read.mtp()	reads a Minitab file (package **foreign**)
read.octave()	reads an Octave file (package **foreign**)
read.spss()	reads an SPSS file (package **foreign**)
read.table()	reads a tabular file (e.g., CSV)
read.xls()	reads an Excel file (package **gdata**)
read.xport()	reads an SAS file (package **foreign**)
readLines()	reads lines from text file
read_excel()	reads an Excel file (package **readxl**)
readxl	package to read Excel files
renderPlot()	renders a reactive plot (package **shiny**)
rep()	function that replicates elements of vectors and lists
reticulate	package to interface with Python language
return()	returns an item from a function
rjson	package for input and output of JSON objects
rnorm()	creates normal distribution random samples
round()	rounds the first argument values
runApp()	runs a shiny application
runif()	creates real value uniform random samples
sample()	creates integer uniform random samples
sapply()	applies a function over a vector

save()	saves an object into a file
save.image()	saves workspace
selectInput()	selects list input control (package **shiny**)
seq()	creates a regular sequence
set.seed()	sets the random generation number (used by **sample**, **runif**, ...)
setwd()	sets the working directory
shiny	package for interactive web application
shinyApp()	calls a shiny application (package **shiny**)
sidebarLayout()	creates a layout (package **shiny**)
sidebarPanel()	creates a sidebar (package **shiny**)
sin()	sine trigonometric function
sink()	diverts output to a file connection
sliderInput()	selects a numeric value (package **shiny**)
sort()	sorts a vector
source()	executes R code from a file
source_python()	executes a Python file (package **reticulate**)
sqrt()	square root of a number
stopCluster()	stops a parallel cluster (package **parallel**)
str()	shows internal structure of object
sum()	sum of all values
summary()	shows a summary of the object
switch()	conditional control function
t.test()	performs a t-student test
tan()	tangent trigonometric function
tic()	starts a timer (package **tictoc**)
tictoc	package for timing R scripts
tiff()	sets graphical device to tiff file
titlePanel()	sets panel application title (package **shiny**)
tm	package for text mining
toJSON()	saves JSON file (package **rjson**)
toc()	computes time elapsed (package **tictoc**)
which()	returns the indexes of an object that follows a logical condition
which.max()	returns the indexes of the maximum value
which.min()	returns the indexes of the minimum value
while()	loops execution command
wilcox.test()	Wilcoxon test
wireframe()	draws a 3D scatter plot (package **lattice**)
write.table()	writes object into tabular file
writeLines()	writes lines into a text file
use_python()	configures which version of Python to use (package **reticulate**)

2.7 Exercises

2.1 In the R environment console, create a vector `v` with 10 elements, all set to 0. Using a single command, replace the indexes 3, 7, and 9 values of `v` with 1.

2.2 Create a vector `v` with all even numbers between 1 and 50.

2.3 Create a matrix `m` of size 3×4, such that:

1. The first row contains the sequence 1, 2, 3, 4.
2. The second row contains the square root of the first row.
3. The third row is computed after step 2 and contains the square root of the second row.
4. The forth column is computed after step 3 and contains the squared values of the third column.

Then, show the matrix values and its row and column sums (use `apply` function) with a precision of two decimal digits.

2.4 Create a function `counteven(x)` that counts how many even numbers are included in a vector `x`, using three approaches:

1. Use a `for()` cycle with an `if()` condition.
2. Use `sapply()` function.
3. Use a condition that is applied directly to `x` (without `if`).

Test the function over the object `x=1:10`.

2.5 Write in a file `maxsin.R` the full R code that is needed to create the `maxsin.pdf` PDF file that appears in top right plot of Fig. 1.5 (Sect. 1.8 and Equation 1.2 explain how **max sin** is defined). Execute the R source file and check if the PDF file is identical to the top right plot of Fig. 1.5.

2.6 Forest fires data exercise:

1. If needed, install and load the `RCurl` package.
2. Use the `getURL` and `write` functions to write the forest fires data (Cortez and Morais, 2007) http://archive.ics.uci.edu/ml/machine-learning-databases/forest-fires/forestfires.csv into a local file.
3. Load the local CSV file (`forestfires.csv`, the separator character is `","`) into a data frame.
4. Show the average temperature in August.
5. Select 10 random samples of temperatures from the months February, July, and August; then check if the average temperature differences are significant under 95% confidence level t-student paired tests.
6. Show all records from August and with a burned area higher than 100.
7. Save the records obtained previously (6) into a CSV file named `aug100.csv`.

Chapter 3
Blind Search

3.1 Introduction

Full blind search assumes the exhaustion of all alternatives, where any previous search does not affect how next solutions are tested (left of Fig. 3.1). Given that the full search space is tested, the optimum solution is always found. Full blind search is only applicable to discrete search spaces, and it is easy to encode in two ways: first, by setting the full search space in a matrix and then sequentially testing each row (solution) of this matrix; and second, in a recursive way, by setting the search space as a tree, where each branch denotes a possible value for a given variable and all solutions appear at the leaves (at the same level). Examples of two known blind methods based on tree structures are the depth-first and breadth-first algorithms. The former starts at the root of the tree and traverses through each branch as far as possible, before backtracking. The latter also starts at the root but searches on a level basis, searching first all succeeding nodes of the root, then the next succeeding nodes of the root succeeding nodes, and so on.

The major disadvantage of pure blind search is that it is not feasible when the search space is continuous or too large, a situation that often occurs with real-world tasks. Consider, for instance, the **bag prices** toy problem defined in Sect. 1.8, even with a small search dimension ($D = 5$), the full search space is quite large for the R tool (i.e., $1000^5 = 10^{15} = 1000 \times 10^{12} = 1000$ billion of searches!). Hence, pure blind search methods are often adapted by setting thresholds (e.g., depth-first search with a maximum depth of K), reducing the space searched or using heuristics. Grid search (Hsu et al., 2003) is an example of a search space reduction method. Monte Carlo search (Caflisch, 1998), also known as random search, is another popular blind method. The method is based on a repeated random sampling, with up to N sampled points. This method is popular since it is computationally feasible and quite easy to encode.

P. Cortez, *Modern Optimization with R*, Use R!,
https://doi.org/10.1007/978-3-030-72819-9_3

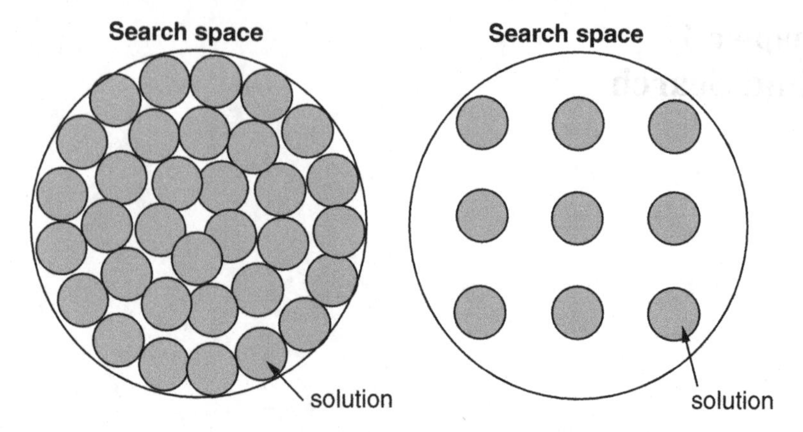

Fig. 3.1 Example of pure blind search (left) and grid search (right) strategies

The next sections present R implementations of three blind search methods: full blind search, grid search, and Monte Carlo search. Also, these implementations are tested on the demonstrative problems presented in Chap. 1.

3.2 Full Blind Search

This section presents two blind search functions: **fsearch** and **dfsearch**. The former is a simpler function that requires the search space to be explicitly defined in a matrix in the format *solutions* $\times D$ (argument **search**), while the latter performs a recursive implementation of the depth-first search and requires the definition of the domain values for each variable to be optimized (argument **domain**). Both functions receive as arguments the evaluation function (**fn**), the optimization type (**type**, a character with **"min"** or **"max"**), and extra arguments (denoted by **...** and that might be used by the evaluation function **fn**). In this chapter, the function arguments (e.g., **fn**, **lower**, **upper**) were named in the same way as they are used by the R **optim** function, described in Sect. 4.3. The **fsearch** and **dfsearch** functions are implemented in file **blind.R**:

```
### blind.R file ###

# full bind search method
#    search - matrix with solutions x D
#    fn - evaluation function
#    type - "min" or "max"
#    ... - extra parameters for fn
fsearch=function(search,fn,type="min",...)
{
 x=apply(search,1,fn,...) # run fn over all search rows
 ib=switch(type,min=which.min(x),max=which.max(x))
```

```
 return(list(index=ib,sol=search[ib,],eval=x[ib]))
}

# depth-first full search method
#    l - level of the tree
#    b - branch of the tree
#    domain - vector list of size D with domain values
#    fn - eval function
#    type - "min" or "max"
#    D - dimension (number of variables)
#    par - current parameters of solution vector
#    bcur - current best sol
#    ... - extra parameters for fn
dfsearch=function(l=1,b=1,domain,fn,type="min",D=length(domain),
                  par=rep(NA,D),
                  bcur=switch(type,min=list(sol=NULL,eval=Inf),
                                   max=list(sol=NULL,eval=-Inf)),
                  ...)
{ if((l-1)==D) # "leave" with solution par to be tested:
     { f=fn(par,...);fb=bcur$eval
       ib=switch(type,min=which.min(c(fb,f)),
                      max=which.max(c(fb,f)))
       if(ib==1) return (bcur) else return(list(sol=par,eval=f))
     }
  else # go through sub branches
     { for(j in 1:length(domain[[l]]))
          { par[l]=domain[[l]][j]
            bcur=dfsearch(l+1,j,domain,fn,type,D=D,
                          par=par,bcur=bcur,...)
          }
       return(bcur)
     }
}
```

The recursive function **dfsearch** tests if the tree node is a leave, computing the evaluation function for the respective solution, else it traverses through the node sub-branches. This function requires some memory state variables (**l**, **b**, **x**, and **bcur**) that are changed each time a new recursive call is executed. The domain of values is stored in a vector list of length D, since the elements of this vector can have different lengths, according to their domain values.

The next R code tests both blind search functions for the **sum of bits** and **max sin** tasks (Sect. 1.8, $D = 8$):

```
### binary-blind.R file ###

source("blind.R") # load the blind search methods

# read D bits from integer x:
binint=function(x,D)
{ x=rev(intToBits(x)[1:D]) # get D bits
  # remove extra 0s from raw type:
  as.numeric(unlist(strsplit(as.character(x),""))[(1:D)*2])
}
```

```
# convert binary vector into integer: code inspired in
# http://stackoverflow.com/questions/12892348/
# in-r-how-to-convert-binary-string-to-binary-or-decimal-value
intbin=function(x) sum(2^(which(rev(x==1))-1))

# sum a raw binary object x (evaluation function):
sumbin=function(x) sum(as.numeric(x))

# max sin of binary raw object x (evaluation function):
maxsin=function(x,Dim) sin(pi*(intbin(x))/(2^Dim))

D=8 # number of dimensions
x=0:(2^D-1) # integer search space
# set full search space in solutions x D:
search=t(sapply(x,binint,D=D))
# set the domain values (D binary variables):
domain=vector("list",D)
for(i in 1:D) domain[[i]]=c(0,1) # bits

# sum of bits, fsearch:
S1=fsearch(search,sumbin,"max") # full search
cat("fsearch best s:",S1$sol,"f:",S1$eval,"\n")

# sum of bits, dfsearch:
S2=dfsearch(domain=domain,fn=sumbin,type="max")
cat("dfsearch best s:",S2$sol,"f:",S2$eval,"\n")

# max sin, fsearch:
Dim=length(search[1,]) # set Dim argument of maxim
S3=fsearch(search,maxsin,"max",Dim=Dim) # Dim used by maxsin
cat("fsearch best s:",S3$sol,"f:",S3$eval,"\n")

# max sin, dfsearch: Dim is used by maxin
S4=dfsearch(domain=domain,fn=maxsin,type="max",Dim=Dim)
cat("dfsearch best s:",S4$sol,"f:",S4$eval,"\n")
```

In the code, **binint** is an auxiliary function that selects only D bits from the **raw** object returned by **intToBits()**. The **intToBits** returns 32 bits in a reversed format, and thus the **rev** R function is also applied to set correctly the order of bits. Given that the **raw** type includes two hex digits, the purpose of the last line of function **binint** is to remove extra 0 characters from the raw object. Such line uses some R functions that were not described in the previous chapter: **as.character**—converts object into character type, **strsplit**—splits a character vector into substrings, and **unlist**—transforms a list into a vector. The following R session exemplifies the effect of the **binint** code (and newly introduced R functions):

```
> x=intToBits(7)[1:4]; print(x)
[1] 01 01 01 00
> x=rev(x); print(x)
[1] 00 01 01 01
```

```
> x=strsplit(as.character(x),""); print(x)
[[1]]
[1] "0" "0"

[[2]]
[1] "0" "1"

[[3]]
[1] "0" "1"

[[4]]
[1] "0" "1"

> x=unlist(x); print(x)
[1] "0" "0" "0" "1" "0" "1" "0" "1"
> x=as.numeric(x[(1:4)*2]); print(x)
[1] 0 1 1 1
```

The generic **sapply** function uses the defined **binint** function in order to create the full binary search space from an integer space. Given that **sapply** returns a $D\times$ *solutions* matrix, the **t** R function is used to transpose the matrix into the required *solutions* $\times D$ format. The result of executing file **binary-blind.R** is:

```
> source("binary-blind.R")
fsearch best s: 1 1 1 1 1 1 1 1 f: 8
dfsearch best s: 1 1 1 1 1 1 1 1 f: 8
fsearch best s: 1 0 0 0 0 0 0 0 f: 1
dfsearch best s: 1 0 0 0 0 0 0 0 f: 1
```

Both methods (**fsearch** and **dfsearch**) return the optimum **sum of bits** and **max sin** solutions.

Turning to the **bag prices** task (Sect. 1.8), as explained previously, the search of all solutions (1000^5) is not feasible in practical terms. However, using domain knowledge, i.e., the original problem formulation assumes that the price for each bag can be optimized independently of other bag prices, it is easy to get the optimum solution, as shown in file **bag-blind.R**:

```
### bag-blind.R file ###

source("blind.R") # load the blind search methods
source("functions.R") # load profit(), cost() and sales()

# auxiliary function that sets the optimum price for
# one bag type (D), assuming an independent influence of
# a particular price on the remaining bag prices:
ibag=function(D) # D - type of bag
{ x=1:1000 # price for each bag type
  # set search space for one bag:
  search=matrix(ncol=5,nrow=1000)
  search[]=1; search[,D]=x
  S1=fsearch(search,profit,"max")
  S1$sol[D] # best price
```

```
}

# compute the best price for all bag types:
S=sapply(1:5,ibag)
# show the optimum solution:
cat("optimum s:",S,"f:",profit(S),"\n")
```

The result of executing file **bag-blind.R** is:

```
> source("bag-blind.R")
optimum s: 414 404 408 413 395 f: 43899
```

It should be noted that while the original formulation of **bag prices** assumes an independent optimization of each bag price (with an optimum profit of 43899), there are other variations presented in this book where this assumption is not true (see Sects. 5.8 and 6.2).

Given that pure blind search cannot be applied to real value search spaces ($\Re$), no code is shown here for the **sphere** and **rastrigin** tasks. Nevertheless, these two real value optimization tasks are handled in the next two sections.

3.3 Grid Search

Grid search reduces the space of solutions by implementing a regular hyperdimensional search with a given step size. The left of Fig. 3.1 shows an example of a two-dimensional (3×3) grid search. Grid search is particularly used for hyperparameter optimization of machine-learning algorithms, such as neural networks or support vector machines (Cortez, 2010).

There are several grid search variants. Uniform design search (Huang et al., 2007) is similar to the standard grid search method, except that is uses a different type of grid, with lesser search points. Nested grid search is another variant that uses several grid search levels. The first level is used with a large step size, then, a second grid level is applied over the best point, searching over a smaller area and with a lower grid size, and so on. Nested search is not a pure blind method, since it incorporates a greedy heuristic, where the next level search is guided by the result of the current level search.

Depending on the grid step size, grid search is often much faster than pure bind search. Also, depending on the number of levels and initial grid step size, nested search might be much faster than standard grid search, but it also can get stuck more easily on local minima. The main disadvantage of the grid search approach is that it suffers from the curse of dimensionality, i.e., the computational effort complexity is very high when the number of dimensions (variables to optimize) is large. For instance, the standard grid search computational complexity is $O(L^D)$, where L is the number of grid search levels and D the dimension (variables to optimize). If only $L = 3$ levels are considered and with a dimension of $D = 30$, this leads to $3^{30} \approx 206$ billion searches, which is infeasible under the R tool. Thus, grid search is not particularly efficient in several practical applications. There are other disadvantages of grid search methods: there are additional parameters that need to be set (e.g., grid search step, the number of nested levels), and they do not guarantee that the optimum solution is obtained.

The next code implements two functions for the standard grid search method (**gsearch** and **gsearch2**) and one for the nested grid search (**ngsearch**):

```
### grid.R file ###

# standard grid search method (uses fsearch)
#    fn - evaluation function
#    lower - vector with lowest values for each dimension
#    upper - vector with highest values for each dimension
#    step - vector with step size for each dimension D
#    type - "min" or "max"
#    ... - extra parameters for fn
gsearch=function(fn,lower,upper,step,type="min",...)
{ D=length(step) # dimension
  domain=vector("list",D) # domain values
  L=vector(length=D) # auxiliary vector
  for(i in 1:D)
     { domain[[i]]=seq(lower[i],upper[i],by=step[i])
       L[i]=length(domain[[i]])
     }
  LS=prod(L)
  s=matrix(ncol=D,nrow=LS) # set the search space
  for(i in 1:D)
    {
     if(i==1) E=1 else E=E*L[i-1]
     s[,i]=rep(domain[[i]],length.out=LS,each=E)
    }
  fsearch(s,fn,type,...) # best solution
}

# standard grid search method (uses dfsearch)
gsearch2=function(fn,lower,upper,step,type="min",...)
{ D=length(step) # dimension
  domain=vector("list",D) # domain values
  for(i in 1:D) domain[[i]]=seq(lower[i],upper[i],by=step[i])
  dfsearch(domain=domain,fn=fn,type=type,...) # solution
}

# nested grid search method (uses fsearch)
#    levels - number of nested levels
ngsearch=function(fn,lower,upper,levels,step,type,...)
{ stop=FALSE;i=1 # auxiliary objects
  bcur=switch(type,min=list(sol=NULL,eval=Inf),
                   max=list(sol=NULL,eval=-Inf))
  while(!stop) # cycle while stopping criteria is not met
  {
     s=gsearch(fn,lower,upper,step,type,...)
     # if needed, update best current solution:
     if( (type=="min" && s$eval<bcur$eval)||
```

```
        (type=="max" && s$eval>bcur$eval)) bcur=s
    if(i<levels) # update step, lower and upper:
      { step=step/2
        interval=(upper-lower)/4
        lower=sapply(lower,max,s$sol-interval)
        upper=sapply(upper,min,s$sol+interval)
      }
    if(i>=levels || sum((upper-lower)<=step)>0) stop=TRUE
    else i=i+1
  }
  return(bcur) # best solution
}
```

All functions require the setting of the lower and upper bounds (vectors **lower** and **upper**) and the grid step (**step**, numeric vector). The first function uses the **fsearch** function, while the second one uses the recursive blind variant (**dfsearch**), both described in Sect. 3.2. The **gsearch** function contains more code in comparison with **gsearch2**, since it requires setting first the search space. This is achieved by using the useful **rep** function with the **each** argument. For example, **rep(1:2,each=2)** returns the vector: **1 1 2 2**. The second grid search function (**gsearch2**) is simpler to implement, given that it performs a direct call of the depth-first search.

The nested grid function uses a simple cycle that calls **gsearch** and whose maximum number of iterations depends on the **levels** argument. The cycle also stops when the range set by the upper and lower bounds is lower than the step size. The next level search is set around the best solution of the current grid search and with half of the current step size. The lower and upper bounds are changed accordingly, and the **min** and **max** functions are used to avoid setting a search space larger than the original bounds. For some configurations of the **step**, **lower**, and **upper** arguments, this nested function might repeat on the next level the evaluation of solutions that were previously evaluated. For the sake of simplicity, the nested grid code is kept with this handicap, although it could be enhanced by implementing a cache that stores previous tested solutions in memory and only computes the evaluation function for new solutions.

The next code explores the three implemented grid search methods for the **bag prices** task of Sect. 1.8:

```
### bag-grid.R file ###

source("blind.R") # load the blind search methods
source("grid.R") # load the grid search methods
source("functions.R") # load the profit function

# grid search for all bag prices, step of 100$
PTM=proc.time() # start clock
S1=gsearch(profit,rep(1,5),rep(1000,5),rep(100,5),"max")
sec=(proc.time()-PTM)[3] # get seconds elapsed
cat("gsearch best s:",S1$sol,"f:",S1$eval,"time:",sec,"s\n")

# grid search 2 for all bag prices, step of 100$
```

```
PTM=proc.time() # start clock
S2=gsearch2(profit,rep(1,5),rep(1000,5),rep(100,5),"max")
sec=(proc.time()-PTM)[3] # get seconds elapsed
cat("gsearch2 best s:",S2$sol,"f:",S2$eval,"time:",sec,"s\n")

# nested grid with 3 levels and initial step of 500$
PTM=proc.time() # start clock
S3=ngsearch(profit,rep(1,5),rep(1000,5),3,rep(500,5),"max")
sec=(proc.time()-PTM)[3] # get seconds elapsed
cat("ngsearch best s:",S3$sol,"f:",S3$eval,"time:",sec,"s\n")
```

This code includes the **proc.time** R function, which returns the time elapsed (in seconds) by the running process and that is useful for computational effort measurements. The result of executing file **bag-grid.R** is:

```
> source("bag-grid.R")
gsearch best s: 401 401 401 401 501 f: 43142 time: 2.055 s
gsearch2 best s: 401 401 401 401 501 f: 43142 time: 2.035 s
ngsearch best s: 376.375 376.375 376.375 501.375 501.375 f: 42823
     time: 0.028 s
```

Under the tested settings, the pure grid search methods execute 10 searches per dimension, leading to a total of $10^5 = 100,000$ evaluations, achieving the same solution (43142) under a similar computational effort.[1] The nested grid achieves a better solution (42823) under much less evaluations (2 searches per dimension and level, a total of $2^5 \times 3 = 96$ tested solutions).

Regarding the real value optimization tasks (**sphere** and **rastrigin**, Sect. 1.8), these can be handled by grid search methods, provided that the dimension adopted is small. The next code shows an example for $D = 2$ and a range of $[-5.2, 5.2]$ (commonly used within these benchmark functions):

```
### real-grid.R file ###

source("blind.R") # load the blind search methods
source("grid.R") # load the grid search methods

# real value functions: sphere and rastrigin:
sphere=function(x) sum(x^2)
rastrigin=function(x) 10*length(x)+sum(x^2-10*cos(2*pi*x))

cat("sphere:\n") # D=2, easy task
S=gsearch(sphere,rep(-5.2,2),rep(5.2,2),rep(1.1,2),"min")
cat("gsearch s:",S$sol,"f:",S$eval,"\n")
S=ngsearch(sphere,rep(-5.2,2),rep(5.2,2),3,rep(3,2),"min")
cat("ngsearch s:",S$sol,"f:",S$eval,"\n")

cat("rastrigin:\n") # D=2, easy task
S=gsearch(rastrigin,rep(-5.2,2),rep(5.2,2),rep(1.1,2),"min")
```

[1] Slightly different execution times can be achieved by executing distinct runs under the same code and machine.

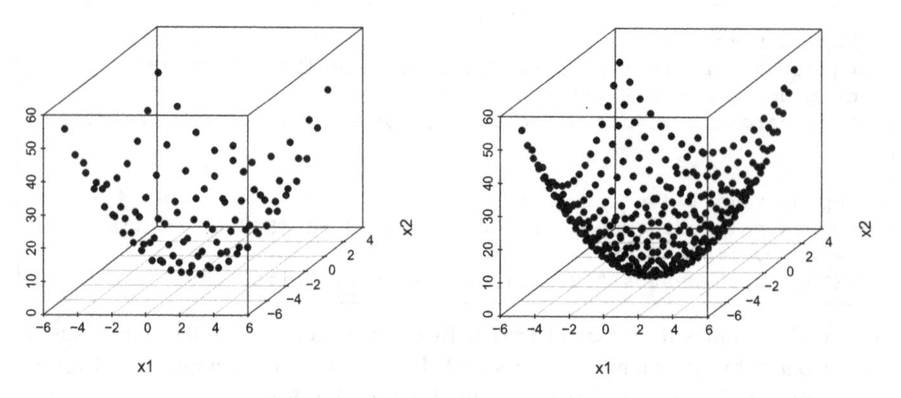

Fig. 3.2 Example of grid search using $L = 10$ (left) and $L = 20$ (right) levels for **sphere** and $D = 2$

```
cat("gsearch s:",S$sol,"f:",S$eval,"\n")
S=ngsearch(rastrigin,rep(-5.2,2),rep(5.2,2),3,rep(3,2),"min")
cat("ngsearch s:",S$sol,"f:",S$eval,"\n")
```

The execution result of file **real-grid.R** is:

```
sphere:
gsearch s: 0.3 0.3 f: 0.18
ngsearch s: -0.1 -0.1 f: 0.02
rastrigin:
gsearch s: -1.9 -1.9 f: 11.03966
ngsearch s: -0.1 -0.1 f: 3.83966
```

Good solutions were achieved, close to the optimum solution of $s = (0, 0)$ and $f = 0$. Figure 3.2 shows how the space is searched when using the standard grid search for **sphere** when using $L = 10$ (left) and $L = 20$ search levels (right) for each dimension. The three-dimensional plots were achieved using the **scatterplot3d** function of the **scatterplot3d** package.

3.4 Monte Carlo Search

Monte Carlo search is a versatile numerical method that is easy to implement and is applicable to high-dimensional problems (in contrast with grid search), ranging from Physics to Finance (Caflisch, 1998). The method consists in a random generation of N points, using a given probability distribution over the problem domain. The computational effort complexity is $O(N)$.

More details about implementing Monte Carlo methods in R can be found in Robert and Casella (2009). In this book, we present a very simple implementation of the Monte Carlo search, which adopts the uniform distribution $\mathcal{U}(lower, upper)$ and includes only four lines of code:

```
### montecarlo.R file ###

# montecarlo uniform search method
#    fn - evaluation function
#    lower - vector with lowest values for each dimension
#    upper - vector with highest values for each dimension
#    N - number of samples
#    type - "min" or "max"
#    ... - extra parameters for fn
mcsearch=function(fn,lower,upper,N,type="min",...)
{ D=length(lower)
  s=matrix(nrow=N,ncol=D) # set the search space
  for(i in 1:N) s[i,]=runif(D,lower,upper)
  fsearch(s,fn,type,...) # best solution
}
```

The proposed implementation is tested here for the **bag prices** ($D = 5$) and real value tasks (**sphere** and **rastrigin**, $D \in \{2, 30\}$) by using $N = 10,000$ uniform samples:

```
### test-mc.R file ###

source("blind.R") # load the blind search methods
source("montecarlo.R") # load the monte carlo method
source("functions.R") # load the profit function

N=10000 # set the number of samples
cat("monte carlo search (N:",N,")\n")

# bag prices
cat("bag prices:")
S=mcsearch(profit,rep(1,5),rep(1000,5),N,"max")
cat("s:",S$sol,"f:",S$eval,"\n")

# real-value functions: sphere and rastrigin:
sphere=function(x) sum(x^2)
rastrigin=function(x) 10*length(x)+sum(x^2-10*cos(2*pi*x))

D=c(2,30)
label="sphere"
for(i in 1:length(D))
   { S=mcsearch(sphere,rep(-5.2,D[i]),rep(5.2,D[i]),N,"min")
     cat(label,"D:",D[i],"s:",S$sol[1:2],"f:",S$eval,"\n")
   }
label="rastrigin"
for(i in 1:length(D))
   { S=mcsearch(rastrigin,rep(-5.2,D[i]),rep(5.2,D[i]),N,"min")
```

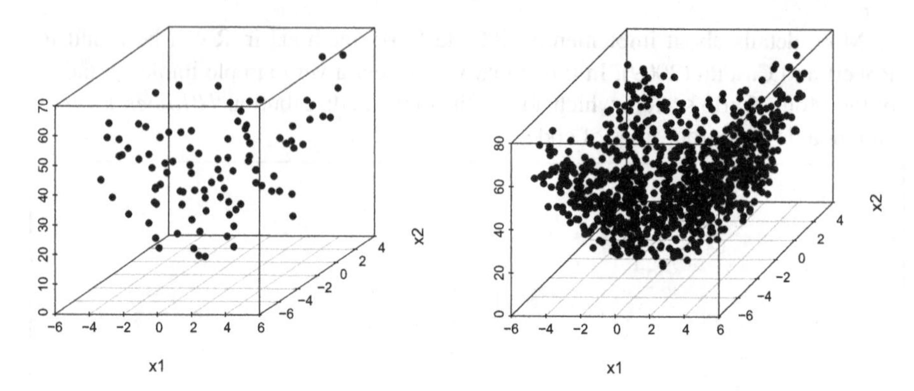

Fig. 3.3 Example of Monte Carlo search using $N = 100$ (left) and $N = 1000$ (right) samples for **sphere** and $D = 2$

```
    cat(label,"D:",D[i],"s:",S$sol[1:2],"f:",S$eval,"\n")
  }
```

To simplify the analysis of the obtained results, the code only shows the optimized values for the first two variables (x_1 and x_2). Given that Monte Carlo is a stochastic method, each run will present a different result. An example execution (one run) is:

```
> source("test-mc.R")
monte carlo search (N: 10000 )
bag prices:s: 349.7477 369.1669 396.1959 320.5007 302.3327 f:
    42508
sphere D: 2 s: -0.01755296 0.0350427 f: 0.001536097
sphere D: 30 s: -0.09818928 -1.883463 f: 113.7578
rastrigin D: 2 s: -0.0124561 0.02947438 f: 0.2026272
rastrigin D: 30 s: 0.6508581 -3.043595 f: 347.1969
```

Under the tested step ($N = 10,000$), interesting results were achieved for the **sphere** and **rastrigin** tasks when $D = 2$. However, when the dimension increases ($D = 30$), the optimized solutions are further away from the optimum value ($f = 0$). For demonstration purposes, Fig. 3.3 shows two examples of Monte Carlo searches for **sphere** and $D = 2$ (left plot with $N = 100$ and right graph with $N = 1000$).

3.5 Command Summary

`as.character()`	converts object into character
`as.numeric()`	converts object into numeric
`dfsearch()`	depth-first blind search (chapter file `"blind.R"`)
`fsearch()`	full blind search (chapter file `"blind.R"`)
`gsearch()`	grid search (chapter file `"grid.R"`)
`intToBits()`	converts integer to binary representation
`max()`	maximum of all values
`mcsearch()`	Monte Carlo search (chapter file `"montecarlo.R"`)
`min()`	minimum of all values
`ngsearch()`	nested grid search (chapter file `"grid.R"`)
`proc.time()`	time elapsed by the running process
`rep()`	function that replicates elements of vectors and lists
`rev()`	reversed version of an object
`scatterplot3d`	package that implements `scatterplot3d()`
`scatterplot3d()`	plots a 3D point cloud (package `scatterplot3d`)
`strsplit()`	splits elements of character vector
`t()`	matrix transpose
`unlist()`	converts list to vector

3.6 Exercises

3.1 Explore the optimization of the binary **max sin** task with a higher dimension ($D = 16$), under pure blind, grid, and Monte Carlo methods. Show the optimized solutions, evaluation values, and time elapsed (in seconds). For the grid and Monte Carlo methods, use directly the `fsearch` function, by changing only the integer space (object `x` of the file `"binary-blind.R"`) with a maximum of $N = 1000$ searches. Tip: use `seq()` for setting the grid search integer space and `sample()` for Monte Carlo.

3.2 Consider the **bag prices** ($D = 5$). Adapt the file `"bag-grid.R"` such that two grid searches are performed over the range [350, 450] with a step size of 11$ in order to show the solution and evaluation values. The first search should be executed using `gsearch` function, while the second search should be implemented using the depth-first method (`dfsearch` function). Tip: use the `seq` function to set the domain of values used by `dfsearch`.

3.3 Consider the **rastrigin** task with a dimension of $D = 30$. Using the Monte Carlo method, explore different N values within the range $\{100,1000,10000\}$. Execute 30 runs for each N value and compare if the average differences are statistically significant at the 95% confidence level under a pairwise t-student test. Also, plot the boxplots for the results due to each N value.

Chapter 4
Local Search

4.1 Introduction

In contrast with the blind search methods presented in Chap. 3, modern optimization techniques are based on a guided search, where new solutions are generated from existing solutions. Singe-state search methods assume just one initial solution that is improved through a series of iterations. Most single-state methods perform a local search, focusing their attention within a local neighborhood of a given initial solution, as shown in Fig. 4.1. This chapter is devoted to such methods. A priori knowledge, such as problem domain heuristics, can be used to set the initial solution. A more common approach is to set the initial point randomly and perform several restarts (also known as runs).

The main differences within local methods are set on how new solutions are defined and what is kept as the current solution (corresponding to functions *change* and *select* of Algorithm 1). The next sections describe how single-state search methods (e.g., hill climbing, simulated annealing, and tabu search) can be adopted in the R tool. This chapter also includes a section that describes how to compare search methods in R, by providing a demonstrative example that compares two local search methods with random search. Finally, the last section of this chapter explains how modern optimization methods (including local search ones) can be tuned in terms of their control parameters.

4.2 Hill Climbing

Hill climbing is a simple local optimization method that "climbs" up the hill until a local optimum is found (assuming a maximization goal). The method works by iteratively searching for new solutions within the neighborhood of current solution, adopting new solutions if they are better, as shown in the pseudo-code

P. Cortez, *Modern Optimization with R*, Use R!,
https://doi.org/10.1007/978-3-030-72819-9_4

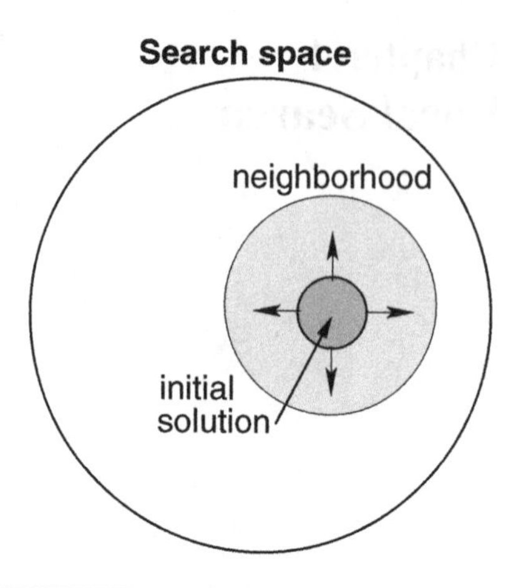

Fig. 4.1 Example of a local search strategy

Algorithm 2 Pure hill climbing optimization method

1: **Inputs:** S, f, C ▷ S is the initial solution, f is the evaluation function, C includes control parameters
2: $i \leftarrow 0$ ▷ i is the number of iterations of the method
3: **while** not $termination\_criteria(S, f, C, i)$ **do**
4: $S' \leftarrow change(S, C)$ ▷ new solution
5: $B \leftarrow best(S, S', f)$ ▷ best solution for next iteration
6: $S \leftarrow B$ ▷ deterministic select function
7: $i \leftarrow i + 1$
8: **end while**
9: **Output:** B ▷ the best solution

of Algorithm 2. The purpose of function *change* is to produce a slightly different solution, by performing a full search in the whole neighborhood or by applying a small random change in the current solution values. It should be noted that while the standard hill climbing algorithm is deterministic, when random changes are used for perturbing a solution, a stochastic behavior is achieved. This is why hill climbing is set at the middle of the deterministic/stochastic dimension in Fig. 1.4.

The R implementation of the standard hill climbing method is coded in file **hill.R**:

```
### hill.R file ###

# pure hill climbing:
#    par - initial solution
#    fn - evaluation function
#    change - function to generate the next candidate
#    lower - vector with lowest values for each dimension
#    upper - vector with highest values for each dimension
#    control - list with stopping and monitoring method:
#       $maxit - maximum number of iterations
#       $REPORT - frequency of monitoring information
#       $digits - (optional) round digits for reporting
#    type - "min" or "max"
#    ... - extra parameters for fn
```

```
hclimbing=function(par,fn,change,lower,upper,control,
                   type="min",...)
{ fpar=fn(par,...)
  for(i in 1:control$maxit)
     {
      par1=change(par,lower,upper)
      fpar1=fn(par1,...)
      if(control$REPORT>0 &&(i==1||i%%control$REPORT==0))
        report_iter(i,par,fpar,par1,fpar1,control)
      b=best(par,fpar,par1,fpar1,type) # select best
      par=b$par;fpar=b$fpar
     }
  if(control$REPORT>=1)
     report_iter("best:",par,fpar,control=control)
  return(list(sol=par,eval=fpar))
}

# report iteration details:
# i, par, fpar, par1 and fpar1 are set in the search method
# control - list with optional number of digits: $digits
report_iter=function(i,par,fpar,par1=NULL,fpar1=NULL,control)
{
 if(is.null(control$digits)) digits=2 # default value
 else digits=control$digits
 if(i=="best:") cat(i,round(par,digits=digits),"f:",round(fpar,
     digits=digits),"\n")
 else cat("i:",i,"s:",round(par,digits=digits),"f:",round(fpar,
     digits=digits),
          "s'",round(par1,digits=digits),"f:",round(fpar1,digits=
              digits),"\n")
}

# slight random change of vector par:
#    par - initial solution
#    lower - vector with lowest values for each dimension
#    upper - vector with highest values for each dimension
#    dist - random distribution function
#    round - use integer (TRUE) or continuous (FALSE) search
#    ... - extra parameters for dist
#    examples: dist=rnorm, mean=0, sd=1; dist=runif, min=0,max=1
hchange=function(par,lower,upper,dist=rnorm,round=TRUE,...)
{ D=length(par) # dimension
  step=dist(D,...) # slight step
  if(round) step=round(step)
  par1=par+step
  # return par1 within [lower,upper]:
  return(ifelse(par1<lower,lower,ifelse(par1>upper,upper,par1)))
}

# return best solution and its evaluation function value
#    par - first solution
#    fpar - first solution evaluation
#    par2 - second solution
#    fpar2 - second solution evaluation
```

```
#    type - "min" or "max"
#    fn - evaluation function
#    ... - extra parameters for fn
best=function(par,fpar,par2,fpar2,type="min",...)
{ if(    (type=="min" && fpar2<fpar)
     || (type=="max" && fpar2>fpar)) { par=par2;fpar=fpar2 }
  return(list(par=par,fpar=fpar))
}
```

The main function (**hclimbing**) receives an initial search point (**par**), an evaluation function (**fn**), **lower** and **upper** bounds, a **control** object, and optimization **type**. The **control** list is used to set the maximum number of iterations (**control$maxit**) and monitor the search, showing the solutions searched at every **control$REPORT** iteration. The optional **control$digits** allows to simplify the visualization of the reported evaluation values (default is **control$digits=2**).

The **report_iter** is a simple function to report the iteration details in the R console. This function uses the **round**, **is.null**, and **cat** functions explained in Chap. 2. The change function (**hchange**) produces a small perturbation over a given solution (**par**). New values are achieved by adopting a random distribution function (**dist**). Given the goal of producing a small perturbation, the normal (Gaussian) distribution $\mathcal{N}(0, 1)$ is adopted in this book, corresponding to the arguments **dist=rnorm, mean=0,** and **sd=1**. This means that in most cases, very small changes are performed (with an average of zero), although large deviations might occur in a few cases. The new solution is kept within the range [**lower**,**upper**] by using the useful **ifelse(***condition*, *yes*, *no***)** R function that performs a conditional element selection (returns the values of *yes* if the *condition* is true, else returns the elements of *no*). For example, the result of **x=c(-1,4,9);sqrt(ifelse(x>=0,x,NA))** is **NA 2 3**. Finally, the **best** function returns a list with the best solution and evaluation value when comparing two solutions.

For demonstration purposes, the next R code executes 10 iterations of a hill climbing search for the **sum of bits** task (Sect. 1.8), starting from the origin (all zero) solution:

```
### sumbits-hill.R file ###

source("hill.R") # load the hill climbing methods

# sum a raw binary object x (evaluation function):
sumbin=function(x) sum(x)

# hill climbing for sum of bits, one run:
D=8 # dimension
s=rep(0,D) # c(0,0,0,0,...)
C=list(maxit=10,REPORT=1) # maximum of 10 iterations
ichange=function(par,lower,upper) # integer change
{ hchange(par,lower,upper,rnorm,mean=0,sd=1) }

hclimbing(s,sumbin,change=ichange,lower=rep(0,D),upper=rep(1,D),
```

```
          control=C,type="max")
```

One example of such execution is:

```
> source("sumbits-hill.R")
i: 1 s: 0 0 0 0 0 0 0 0 f: 0 s' 0 0 0 1 0 0 1 0 f: 2
i: 2 s: 0 0 0 1 0 0 1 0 f: 2 s' 0 0 0 1 0 0 1 0 f: 2
i: 3 s: 0 0 0 1 0 0 1 0 f: 2 s' 0 0 0 0 1 1 0 0 f: 2
i: 4 s: 0 0 0 1 0 0 1 0 f: 2 s' 1 0 0 1 0 0 1 0 f: 3
i: 5 s: 1 0 0 1 0 0 1 0 f: 3 s' 0 0 0 0 1 0 0 1 f: 2
i: 6 s: 1 0 0 1 0 0 1 0 f: 3 s' 1 1 0 1 1 0 0 1 f: 5
i: 7 s: 1 1 0 1 1 0 0 1 f: 5 s' 0 1 0 1 0 1 0 0 f: 3
i: 8 s: 1 1 0 1 1 0 0 1 f: 5 s' 0 0 0 1 0 1 1 0 f: 3
i: 9 s: 1 1 0 1 1 0 0 1 f: 5 s' 1 0 1 1 1 0 1 0 f: 5
i: 10 s: 1 1 0 1 1 0 0 1 f: 5 s' 1 1 0 1 1 1 1 1 f: 7
best: 1 1 0 1 1 1 1 1 f: 7
```

The **sum of bits** is an easy task, and after 10 iterations, the hill climbing method achieves a solution that is very close to the optimum ($f = 8$).

The next code performs a hill climbing for the **bag prices** ($D = 5$) and **sphere** tasks ($D = 2$):

```
### bs-hill.R file ###

source("hill.R") # load the hill climbing methods
source("functions.R") # load the profit function

# hill climbing for all bag prices, one run:
D=5; C=list(maxit=10000,REPORT=10000) # 10000 iterations
s=sample(1:1000,D,replace=TRUE) # initial search
ichange=function(par,lower,upper) # integer value change
{ hchange(par,lower,upper,rnorm,mean=0,sd=1) }
hclimbing(s,profit,change=ichange,lower=rep(1,D),
          upper=rep(1000,D),control=C,type="max")

# hill climbing for sphere, one run:
sphere=function(x) sum(x^2)
D=2; C=list(maxit=10000,REPORT=10000,digits=3)
rchange=function(par,lower,upper) # real value change
{ hchange(par,lower,upper,rnorm,mean=0,sd=0.5,round=FALSE) }

s=runif(D,-5.2,5.2) # initial search
hclimbing(s,sphere,change=rchange,lower=rep(-5.2,D),
          upper=rep(5.2,D),control=C,type="min")
```

An execution example is:

```
> source("bs-hill.R")
i: 1 s: 714 347 926 673 107 f: 27380 s' 715 345 925 673 106 f:
    27293
i: 10000 s: 732 351 943 700 129 f: 28462 s' 730 352 943 699 128 f
    : 28104
best: 732 351 943 700 129 f: 28462
i: 1 s: -2.634 -2.501 f: 13.192 s' -2.802 -3.174 f: 17.927
```

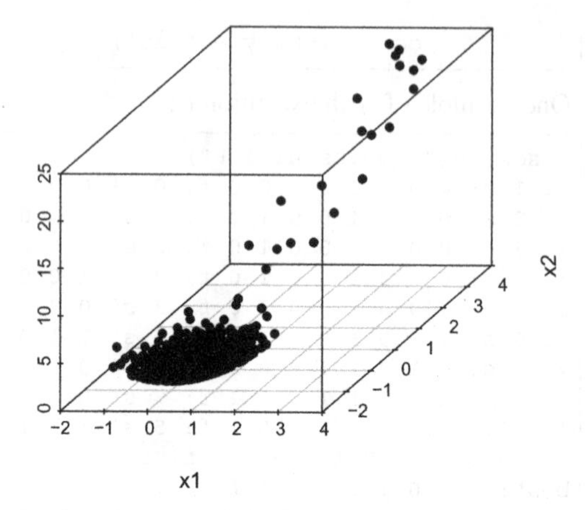

Fig. 4.2 Example of hill climbing search (only best "down the hill" points are shown) for **sphere** and $D = 2$

```
i: 10000 s: 0.012 -0.001 f: 0 s' 0.692 -0.606 f: 0.846
best: 0.012 -0.001 f: 0
```

Using 10,000 iterations, the hill climbing search improved the solution from 27,293 to 28,104 (**bag prices**) and from 17.927 to 0.000 (**sphere**). For demonstrative purposes, Fig. 4.2 shows an example of the searched "down the hill" (best) points for the **sphere** task.

There are several hill climbing variants, such as the steepest ascent hill climbing (Luke, 2015), which searches for up to N solutions in the neighborhood of S and then adopts the best one, and stochastic hill climbing (Michalewicz et al., 2006), which replaces the deterministic select function, selecting new solutions with a probability of P (a similar strategy is performed by the simulated annealing method, discussed in the next section). The steepest ascent executes a Monte Carlo search (Sect. 3.4) around the neighborhood of the current solution, performing N random searches before "climbing the hill." The stochastic hill climbing allows to accept inferior solutions with a probability P, and thus it can escape local minima. In Fernandes et al. (2015), this stochastic variant was successfully adopted to search for adaptations (e.g., new keywords, shorter title size) that could improve the popularity of an online news article (prior to its publication). Several P values were tested, ranging from $P = 0$ (pure hill climbing) to $P = 1$ (Monte Carlo search). Using 1000 news from the Mashable platform (http://mashable.com/), the best results were achieved for $P = 0.2$ (without keyword changes) and $P = 0.8$ (with keyword suggestions).

The two hill climbing variants are implemented in file **hill2.R**:

```
### hill2.R file ###

# steepest ascent hill climbing:
#    par - initial solution
#    fn - evaluation function
#    change - function to generate the next candidate
#    lower - vector with lowest values for each dimension
```

```
#    upper - vector with highest values for each dimension
#    control - list with stopping and monitoring method:
#       $N - number of change searches (steepest ascent)
#       $maxit - maximum number of iterations
#       $REPORT - frequency of monitoring information
#       $digits - (optional) round digits for reporting
#    type - "min" or "max"
#    ... - extra parameters for fn
sa_hclimbing=function(par,fn,change,lower,upper,control,
                      type="min",...)
{ fpar=fn(par,...)
  for(i in 1:control$maxit)
     {
      # first change
      par1=change(par,lower,upper)
      fpar1=fn(par1,...)
      if(control$N>1) # steepest ascent cycle
      { for(j in 1:(control$N-1))
         { # random search for better par1 solutions:
          par2=change(par,lower,upper)
          fpar2=fn(par2,...)
          b=best(par1,fpar1,par2,fpar2,type)
          par1=b$par;fpar1=b$fpar # update change
         }
      }
      if(control$REPORT>0 &&(i==1||i%%control$REPORT==0))
         report_iter(i,par,fpar,par1,fpar1,control)
      b=best(par,fpar,par1,fpar1,type) # update best solution
      par=b$par;fpar=b$fpar
     }
  if(control$REPORT>=1)
     report_iter("best:",par,fpar,control=control)
  return(list(sol=par,eval=fpar))
}

# stochastic hill climbing:
#    par - initial solution
#    fn - evaluation function
#    change - function to generate the next candidate
#    lower - vector with lowest values for each dimension
#    upper - vector with highest values for each dimension
#    control - list with stopping and monitoring method:
#       $P - Probability (in [0,1]) for accepting solutions
#       $maxit - maximum number of iterations
#       $REPORT - frequency of monitoring information
#       $digits - (optional) round digits for reporting
#    type - "min" or "max"
#    ... - extra parameters for fn
st_hclimbing=function(par,fn,change,lower,upper,control,
                      type="min",...)
{ fpar=fn(par,...)
  b=list(par=par,fpar=fpar) # set initial best
  for(i in 1:control$maxit)
     {
```

```
      par1=change(par,lower,upper)
      fpar1=fn(par1,...)
      if(control$REPORT>0 &&(i==1||i%%control$REPORT==0))
         report_iter(i,par,fpar,par1,fpar1,control)

      b=best(b$par,b$fpar,par1,fpar1,type) # memorize best

      x=runif(1) # random between [0,1]
      if(x<control$P) # accept new solution
       { par=par1; fpar=fpar1 }
      else # select best between par and par1
       {
        b1=best(par,fpar,par1,fpar1,type)
        par=b1$par;fpar=b1$fpar # update par
       }
    }
  par=b$par;fpar=b$fpar # set par to best
  if(control$REPORT>=1)
     report_iter("best:",par,fpar,control=control)
  return(list(sol=par,eval=fpar))
}
```

The functions **sa_hclimbing** and **st_hclimbing** implement the steepest ascent and stochastic variants. The method parameters (N and P) are defined in the control list argument: **control$N** for **sa_hclimbing** and **control$P** for **st_hclimbing**. The two functions are demonstrated with the **max sin** task in file **maxsin-hill.R**:

```
### maxsin-hill.R file ###

# load the hill climbing methods
source("hill.R")
source("hill2.R")

intbin=function(x) # convert binary to integer
{ sum(2^(which(rev(x==1))-1)) } # explained in Chapter 3
# max sin of binary raw object x (evaluation function):
maxsin=function(x,Dim) sin(pi*(intbin(x))/(2^Dim))

# hill climbing variants for max sin:
D=8 # dimension
# initial solution:
s=rep(0,D) # c(0,0,0,0,...)
C=list(maxit=10,REPORT=1) # maximum of 10 iterations
lower=rep(0,D); upper=rep(1,D)
ichange=function(par,lower,upper) # integer change
{ hchange(par,lower,upper,rnorm,mean=0,sd=1) }

set.seed(123) # set for replicability

# steepest ascent with 5 searches in each iteration:
C1=C;C1$N=5
cat("steepest ascent: N=",C1$N,"\n")
```

```
sa_hclimbing(s,maxsin,change=ichange,lower=lower,upper=upper,
            control=C1,type="max",Dim=D) # Dim used by maxsin

# low probability schochastic hill climbing:
C2=C;C2$P=0.2
cat("stochastic hill climbing: P=",C2$P,"\n")
st_hclimbing(s,maxsin,change=ichange,lower=lower,upper=upper,
            control=C2,type="max",Dim=D) # Dim used by maxsin
```

In the demonstration, the **max sin** task was set with a dimension of **`D=8`** bits. The **`set.seed`** function was used to generate the same random numbers, and thus the code will produce the same result in different computer machines. In both cases, the methods are executed with **`maxit=10`** iterations. The steepest ascent uses $N = 5$ and the stochastic hill climbing a value of $P = 0.2$. The execution result is:

```
> source("maxsin-hill.R")
steepest ascent: N= 5
i: 1 s: 0 0 0 0 0 0 0 0 f: 0 s' 0 0 1 0 0 1 0 0 f: 0.43
i: 2 s: 0 0 1 0 0 1 0 0 f: 0.43 s' 1 0 0 1 0 0 0 0 f: 0.98
i: 3 s: 1 0 0 1 0 0 0 0 f: 0.98 s' 1 0 0 1 0 0 1 0 f: 0.98
i: 4 s: 1 0 0 1 0 0 0 0 f: 0.98 s' 1 0 0 1 1 0 0 0 f: 0.96
i: 5 s: 1 0 0 1 0 0 0 0 f: 0.98 s' 1 0 0 1 0 0 1 0 f: 0.98
i: 6 s: 1 0 0 1 0 0 0 0 f: 0.98 s' 1 0 0 0 0 0 0 0 f: 1
i: 7 s: 1 0 0 0 0 0 0 0 f: 1 s' 1 0 1 0 0 0 0 0 f: 0.92
i: 8 s: 1 0 0 0 0 0 0 0 f: 1 s' 1 0 0 0 1 1 1 1 f: 0.98
i: 9 s: 1 0 0 0 0 0 0 0 f: 1 s' 1 0 0 0 0 0 0 1 f: 1
i: 10 s: 1 0 0 0 0 0 0 0 f: 1 s' 1 0 0 1 0 1 0 0 f: 0.97
best: 1 0 0 0 0 0 0 0 f: 1
stochastic hill climbing: P= 0.2
i: 1 s: 0 0 0 0 0 0 0 0 f: 0 s' 0 0 0 0 1 0 0 1 f: 0.11
i: 2 s: 0 0 0 0 1 0 0 1 f: 0.11 s' 0 0 0 1 0 0 1 1 f: 0.23
i: 3 s: 0 0 0 1 0 0 1 1 f: 0.23 s' 0 1 0 1 0 0 1 0 f: 0.84
i: 4 s: 0 1 0 1 0 0 1 0 f: 0.84 s' 1 1 0 0 1 0 1 0 f: 0.62
i: 5 s: 1 1 0 0 1 0 1 0 f: 0.62 s' 1 0 1 1 0 1 0 0 f: 0.8
i: 6 s: 1 0 1 1 0 1 0 0 f: 0.8 s' 1 0 0 1 0 1 0 1 f: 0.97
i: 7 s: 1 0 0 1 0 1 0 1 f: 0.97 s' 1 0 1 1 0 1 0 1 f: 0.8
i: 8 s: 1 0 0 1 0 1 0 1 f: 0.97 s' 1 0 1 1 0 1 0 1 f: 0.8
i: 9 s: 1 0 1 1 0 1 0 1 f: 0.8 s' 1 1 1 1 0 1 0 1 f: 0.13
i: 10 s: 1 0 1 1 0 1 0 1 f: 0.8 s' 0 1 0 1 0 0 0 0 f: 0.83
best: 1 0 0 1 0 1 0 1 f: 0.97
```

In the execution, the steepest ascent method has consistently improved the solution, reaching the optimum solution of 1.0. As for the stochastic hill climbing, in 20% of the cases (iterations 4 and 8), it accepted inferior solutions that were passed to the next iteration. Overall, this variant reached a solution that is close to the optimum value (0.97). It should be noted that this execution is shown just for tutorial purposes. In effect, the comparison between the two search methods is not fair, since the steepest ascent performs more solution evaluations (Sect. 4.5 exemplifies how to achieve a proper comparison between local search methods).

4.3 Simulated Annealing

Simulated annealing is a variation of the hill climbing technique that was proposed by Kirkpatrick et al. (1983). The method is inspired in the annealing phenomenon of metallurgy, which involves first heating a particular metal and then performing a controlled cooling (Luke, 2015). This single-state method differs from the hill climbing search by adopting a control temperature parameter (T) that is used to compute the probability of accepting inferior solutions. In contrast with the stochastic hill climbing, which adopts a fixed value for the probability (P), the simulated annealing uses a variable temperature value during the search. The method starts with a high temperature and then gradually decreases (cooling process) the control parameter until a small value is achieved (similar to the hill climbing). Given that simulated annealing is a single-state method, it is described in this chapter. However, it should be noted that for high temperatures, the method is almost equivalent to the Monte Carlo search, thus behaving more like a global search method (in particular, if the *change* function is set to perform high changes), while for low temperatures the method is similar to the hill climbing local search (Michalewicz et al., 2006).

This section details first the simulated annealing implementation of the **optim** R function, which is a known function that comes with the base R installation (package **stats**). The function only performs minimization tasks and that executes several optimization methods by setting argument **method**, such as

- **"Nelder-Mead"**)—Nelder and Mead or downhill simplex method;
- **"BFGS"**—a quasi-Newton method;
- **"CG"**—conjugate gradient method;
- **"L-BFGS-B"**—modification of the BFGS method with lower and upper bounds; and
- **"SANN"**—simulated annealing.

Algorithm 3 presents the pseudo-code of the simulated annealing implementation, which is based on the variant proposed by Bélisle (1992). This implementation includes three search parameters: $maxit$—the maximum number of iterations, **temp** (T)—the initial temperature, and $tmax$—the number of evaluations at each temperature. By default, the values for the control parameters are $maxit = 10000$, $T = 10$, and $tmax = 10$. Also, new search points are generated using a Gaussian Markov kernel with a scale proportional to the temperature. Nevertheless, these defaults can be changed by setting two **optim** arguments: the **control** list and **gr** (*change*) function. The last argument is useful for solving combinatorial problems, i.e., when the representation of the solution includes discrete values. The **optim** function returns a list with several components, such as **$par**—the optimized values and **$value**—the evaluation of the best solution.

Similarly to the hill climbing demonstration, the **sumbits-sann.R** file executes 10 iterations of the simulated annealing for the **sum of bits** task:

Algorithm 3 Simulated annealing search as implemented by the **optim** function

1: **Inputs:** S, f, C ▷ S is the initial solution, f is the evaluation function, C contains control parameters ($maxit$, T and $tmax$)
2: $maxit \leftarrow get\_maxit(C)$ ▷ maximum number of iterations
3: $T \leftarrow get\_temperature(C)$ ▷ temperature, should be a high number
4: $tmax \leftarrow get\_tmax(C)$ ▷ number of evaluations at each temperature
5: $fs \leftarrow f(S)$ ▷ evaluation of S
6: $B \leftarrow S$ ▷ best solution
7: $i \leftarrow 0$ ▷ i is the number of iterations of the method
8: **while** $i < maxit$ **do** ▷ $maxit$ is the termination criterion
9: **for** $j = 1 \rightarrow tmax$ **do** ▷ cycle j from 1 to $tmax$
10: $S' \leftarrow change(S, C)$ ▷ new solution (might depend on T)
11: $fs' \leftarrow f(S')$ ▷ evaluation of S'
12: $r \leftarrow \mathcal{U}(0, 1)$ ▷ random number, uniform within [0, 1]
13: $p \leftarrow \exp(\frac{fs'-fs}{T})$ ▷ probability $P(S, S', T)$ (Metropolis function)
14: **if** $fs' < fs \vee r < p$ **then** $S \leftarrow S'$ ▷ accept best solution or worst if $r < p$
15: **end if**
16: **if** $fs' < fs$ **then** $B \leftarrow S'$
17: **end if**
18: $i \leftarrow i + 1$
19: **end for**
20: $T \leftarrow \frac{T}{\log(i/tmax) \times tmax + \exp(1)}$ ▷ cooling step (decrease temperature)
21: **end while**
22: **Output:** B ▷ the best solution

```
### sumbits-sann.R file ###
source("hill.R") # get hchange function
# sum a raw binary object x (evaluation function):
minsumbin=function(x) (length(x)-sum(x)) # optim only minimizes!

# SANN for sum of bits, one run:
D=8 # dimension
s=rep(0,D) # c(0,0,0,0,...)
C=list(maxit=10,temp=10,tmax=1,trace=TRUE,REPORT=1)
bchange=function(par) # binary change
{ D=length(par)
  hchange(par,lower=rep(0,D),upper=rep(1,D),rnorm,mean=0,sd=1)
}
s=optim(s,minsumbin,gr=bchange,method="SANN",control=C)
cat("best:",s$par,"f:",s$value,"(max: fs:",sum(s$par),")\n")
```

Given that **optim** only performs minimization, the evaluation function needs to be adapted (as discussed in Sect. 1.3). In this example, it was set to have a minimum of zero. Also, given that **method="SANN"** does not include lower and upper bounds, it is the responsibility of the change function (**gr**) to not generate unfeasible solutions. In this case, the auxiliary binary change function (**bchange**) uses the **hchange** function (from file **hill.R**) to set the 0 and 1 bounds for all D values. The **method="SANN"** implementation is stochastic, and thus each run will produce a (slight) different result. An execution example of file **sumbits-sann.R** is:

```
> source("sumbits-sann.R")
sann objective function values
initial       value 8.000000
iter        1 value 4.000000
iter        2 value 2.000000
iter        3 value 2.000000
iter        4 value 2.000000
iter        5 value 2.000000
iter        6 value 2.000000
iter        7 value 2.000000
iter        8 value 2.000000
iter        9 value 2.000000
final         value 2.000000
sann stopped after 9 iterations
best: 0 1 1 0 1 1 1 1 f: 2 (max: fs: 6 )
```

The simulated annealing search is also adapted for **bag prices** ($D = 5$) and **sphere** tasks ($D = 2$), by setting $maxit = 10000$, $T = 1000$, and $tmax = 10$ (file **bs-sann.R**):

```
### bs-sann.R file ###

source("hill.R") # load the hchange method
source("functions.R") # load the profit function
eval=function(x) -profit(x) # optim minimizes!

# hill climbing for all bag prices, one run:
D=5; C=list(maxit=10000,temp=1000,trace=TRUE,REPORT=10000)
s=sample(1:1000,D,replace=TRUE) # initial search
ichange=function(par) # integer value change
{ D=length(par)
  hchange(par,lower=rep(1,D),upper=rep(1000,D),rnorm,mean=0,sd=1)
}
s=optim(s,eval,gr=ichange,method="SANN",control=C)
cat("best:",s$par,"profit:",abs(s$value),"\n")

# hill climbing for sphere, one run:
sphere=function(x) sum(x^2)
D=2; C=list(maxit=10000,temp=1000,trace=TRUE,REPORT=10000)

s=runif(D,-5.2,5.2) # initial search
# SANN with default change (gr) function:
s=optim(s,sphere,method="SANN",control=C)
cat("best:",s$par,"f:",s$value,"\n")
```

An example execution of file **bs-sann.R** is:

```
> source("bs-sann.R")
sann objective function values
initial       value -35982.000000
final         value -39449.000000
```

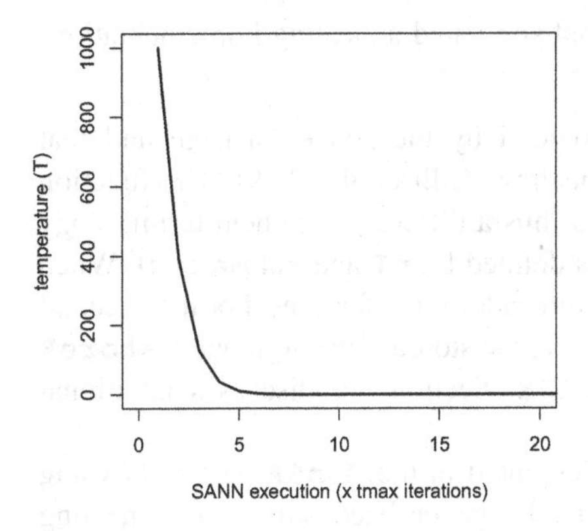

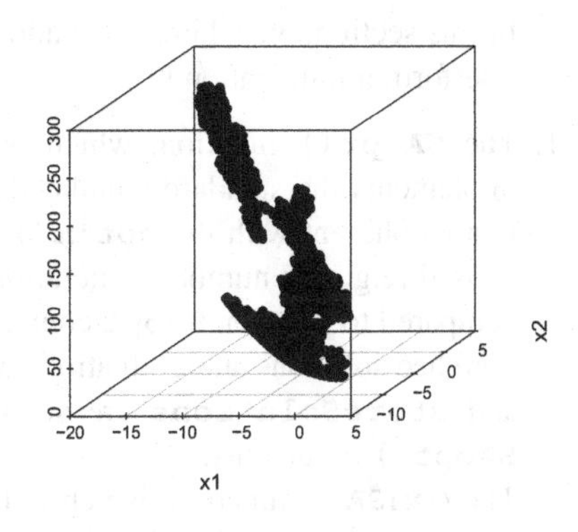

Fig. 4.3 Example of the temperature cooling (left) and simulated annealing search (right) for **sphere** and $D = 2$

```
sann stopped after 9999 iterations
best: 293 570 634 606 474 profit: 39449
sann objective function values
initial        value 21.733662
final          value 1.243649
sann stopped after 9999 iterations
best: -0.6856747 -0.8794882 f: 1.243649
```

In this execution, the initial solution was improved by 3467$ (**bag prices**) and 20.49 (**sphere**). For the last task execution, Fig. 4.3 shows the evolution of the temperature values (left) and points searched (right). While the initial point is within the $[-5.2, 5.2]$ range, the default and unbounded Gaussian Markov *change* function searches for several solutions outside the initial range. However, as the search proceeds, more solutions tend to converge to the origin point, which is the optimum. It is interesting to notice that the left plot of Fig. 4.3 looks like a metal object (e.g., rusty piece), which seems to reflect the metallurgy inspiration of the method.

Similarly to other methods further discussed in this book, there are other simulated annealing implementations provided by independent R packages that need to be installed. There are several advantages of using these packages: often, less code needs to be written; users have access to a richer set of features and flexibility to execute the optimization task. However, it should be noted that there are also disadvantages. First, the user needs to learn how the package works and how to use it. For instance, distinct packages often use different function names and ways to code the same parameters. Second, dependency on packages can lead to what is known as technical debt, where a short-term benefit in using an externally produced code might result in future software maintenance issues (Alves et al., 2016). For example, an older package used by some optimization task code may not be available in a more recent R version. Thus, using a high number of R packages increases the risk of technical debt.

In this section, we address two additional simulated annealing implementations that perform minimization tasks:

1. The **SAopt()** function, which is provided by the **NMOF** package and that implements the standard simulated annealing (Gilli et al., 2019). The function is not coherent with the **optim()**, and thus a different argument terminology is used (e.g., the number of iterations is defined by **nI** and not **maxit**). When compared to **optim**, **SAopt** contains some additional features. For instance, all searched solutions and evaluation values can be stored if the arguments **storeF** and **storeSolutions** are set to **TRUE**. Section 4.6 discusses additional **SAopt()** arguments.
2. The **GenSA()** function, which is implemented in the **GenSA** package (Xiang et al., 2013). The function implements the generalized simulated annealing (GSA) algorithm that was designed to search for a global minimum in complex evaluation functions with a high number of local minima. GSA uses a specific distorted Cauchy–Lorentz visiting distribution to change the solutions (Tsallis and Stariolo, 1996). The **GenSA()** implementation is coherent with the **optim()** notation. Most of the input arguments are shared with **optim()**. An exception is the initial temperature (T), which in **GenSA()** is set by **temperature**, while **optim()** uses **temp**.

A simple demonstration of the different simulated annealing functions is provided in file **rastrigin-sann.R**:

```
### rastrigin-sann.R file ###

# execute 3 simulation annealing optimizations for rastrigin D=30

library(NMOF) # load NMOF
library(GenSA) # load GenSA

source("hill.R") # get hchange

# real value rastrigin function with a dimension of 30:
rastrigin=function(x) 10*length(x)+sum(x^2-10*cos(2*pi*x))
D=30

set.seed(123) # set for replicability

# common setup for the 3 executions:
lower=rep(-5.2,D);upper=rep(5.2,D) # lower and upper bounds
par=runif(D,lower[1],upper[1]) # initial search point
maxit=10000 # maximum number of iterations
temp=1000 # initial temperature
repit=2000 # report every repit iterations

# control argument for optim:
C1=list(maxit=maxit,temp=temp,trace=TRUE,REPORT=repit/10)
# change function (gr argument):
rchange=function(par) # real change
{ D=length(par)
```

```
  # upper and lower are defined globally
  hchange(par,lower=lower,upper=upper,rnorm,mean=0,sd=1)
}
cat("SANN optim execution:\n")
s1=optim(par,rastrigin,gr=rchange,method="SANN",control=C1)
cat("best:",s1$par,"f:",s1$value,"\n")

# algo configuration:
algo=list(nI=maxit,initT=temp,x0=par,neighbour=rchange,
    printDetail=repit,storeF=TRUE)
cat("SAopt optim execution:\n")
s2=SAopt(rastrigin,algo)
cat("best:",s2$xbest,"f:",s2$OFvalue,"\n")

C3=list(maxit=maxit,temperature=temp,verbose=TRUE)
cat("GenSA optim execution:\n")
s3=GenSA(par,rastrigin,lower=lower,upper=upper,control=C3)
cat("best:",s3$par,"f:",s3$value,"\n")
```

The demonstration uses the **rastrigin** function with a large dimension of $D = 30$. This function has several local minima, and thus it is more difficult to optimize than the simpler **sphere** task. Similarly to the **sphere** execution, a $[-5.2, 5.2]$ search range is adopted, starting with a random initial search point. The three methods were set with the same configuration, where $T = 1000$ and $maxit = 10000$, and the results are reported every 2000 iterations. Since the **SAopt()** function does not have a default *change* function, the same **rchange** function is used by both **optim** and **SAopt()**. The random change (**rchange**) is not used in **GenSA()** because the last function uses a fixed function (Cauchy–Lorentz based) to change solutions. The result of the **rastrigin-sann.R** file execution is:

```
> source("rastrigin-sann.R")
SANN optim execution:
sann objective function values
initial       value 558.178269
iter     2000 value 396.894902
iter     4000 value 364.094902
iter     6000 value 364.094902
iter     8000 value 364.094902
iter     9999 value 364.094902
final         value 364.094902
sann stopped after 9999 iterations
best: 3.2 -1.2 0.8 0.2 1.8 -2.2 3.2 -1.2 0.2 -2.8 1.2 2.2 1.2 1.2
     1.2 -0.2 0.2 -0.8 1.8 4.2 -3.2 -0.2 -1.8 4.2 0.2 0.2 -4.2
    -3.2 4.2 2.8 f: 364.0949
SAopt optim execution:

Simulated Annealing.

Running Simulated Annealing ...
Initial solution:  558.1783
Best solution (iteration 2000/10000): 406.0949
Best solution (iteration 4000/10000): 373.0949
```

```
Best solution (iteration 6000/10000): 340.0949
Best solution (iteration 8000/10000): 340.0949
Best solution (iteration 10000/10000): 340.0949
Finished.
Best solution overall: 340.0949
best: 0.2 -2.2 -1.8 -1.2 1.8 1.8 -1.8 1.2 0.8 2.2 -0.2 -1.2 -0.2
    4.2 -3.2 -1.2 -3.2 0.8 1.2 1.8 -3.8 -0.2 -0.8 -4.2 1.8 -0.2
    -3.8 -0.2 2.2 -2.2 f: 340.0949
GenSA optim execution:
It: 1, obj value: 202.9703972
It: 7, obj value: 88.55114422
It: 9, obj value: 58.7024734
It: 11, obj value: 47.75795908
It: 13, obj value: 42.78316883
It: 14, obj value: 35.81846551
It: 17, obj value: 25.86892541
It: 24, obj value: 15.91934491
It: 42, obj value: 11.93950869
It: 48, obj value: 9.949590571
It: 73, obj value: 8.954631514
It: 76, obj value: 6.9647134
It: 87, obj value: 4.974795285
It: 114, obj value: 3.979836228
It: 117, obj value: 2.984877171
It: 180, obj value: 1.989918114
It: 225, obj value: 0.9949590571
It: 267, obj value: 0
.................................................................
.................................................................
.................................................................
.................................................................
..........................best: -1.11752e-11 3.794916e-12
    -3.131309e-10 1.039877e-10 -3.949417e-10 -5.419647e-12
    -2.474186e-10 -1.191516e-11 2.452172e-10 2.855549e-10
    -4.99921e-12 -1.152797e-10 9.905575e-11 -5.193844e-11 4
    .690682e-10 1.840132e-10 -1.381053e-11 -2.145051e-10
    -2.731142e-12 -1.398343e-11 -3.827828e-10 -1.431268e-10 5
    .709847e-09 3.533545e-10 -1.688546e-11 1.28575e-11 2
    .395883e-12 -3.134943e-10 1.540772e-11 1.018197e-10 f: 0
```

In this execution, intermediate results are reported every 2000 iterations, except for **GenSA**, which automatically selects the intermediate results to show. In this simulation, **optim()** obtained a minimum of 364.1, **SAopt()** achieved a slightly better result (340.1), and **GenSA()** managed to get the global optimum result (0.0). It should be noted that this demonstration uses just a single run, and thus the comparison is not robust (see Sect. 4.5). Nevertheless, the high performance of GSA in this example is consistent with the results shown in Xiang et al. (2013), where **GenSA** compared favorable with **method="sann"** of **optim** for the **rastrigin** task.

4.4 Tabu Search

Tabu search was created by Fred Glover (1986) and uses the concept of "memory" to force the search into new areas. The algorithm is a variation of the hill climbing method that includes a tabu list of length L, which stores the most recent solutions that become "tabu" and thus cannot be used when selecting a new solution. The intention is to keep a short-term memory of recent changes, preventing future moves from deleting these changes (Brownlee, 2011). Similarly to the hill climbing method, the search of solutions within the neighborhood of the current solution (function *change*) can be deterministic, including the entire neighborhood, or stochastic (e.g., small random perturbation). Also, the tabu search algorithm is deterministic, except if a stochastic change is adopted (Michalewicz et al., 2007). Hence, this method is also centered within the deterministic/stochastic factor of analysis in Fig. 1.4.

There are extensions of the original tabu search method. Tabu search was devised for discrete spaces and combinatorial problems (e.g., traveling salesman problem). However, the method can be extended to work with real-valued points if a similarity function is used to check if a solution is very close to a member of the tabu list (Luke, 2015). Other extensions to the original method include adding other types of memory structures, such as intermediate term, to focus the search in promising areas (intensification phase), and long term, to promote a wider exploration of the search space (diversification phase). More details can be found in Glover (1990).

Algorithm 4 presents a simple implementation of tabu search, in an adaptation of the pseudo-code presented in (Brownlee, 2011; Luke, 2015). The algorithm combines a steepest ascent hill climbing search with a short tabu memory and includes three control parameters: $maxit$—the maximum number of iterations, L—the length of the tabu list, and N—the number of neighborhood solutions searched at each iteration.

In this section, the **`tabuSearch`** function is adopted (as implemented in the package under the same name). This function only works with binary strings, using a stochastic generation of new solutions and assuming a maximization goal. Also, it implements a three memory scheme, under the sequence of stages: preliminary search (short term), intensification (intermediate term), and diversification (long term). Some relevant arguments are

- **`size`**—length of the binary solution (L_S);
- **`iters`**—maximum number of iterations ($maxit$) during the preliminary stage;
- **`objFunc`**—evaluation function (f) to be maximized;
- **`config`**—initial solution (S);
- **`neigh`**—number of neighbor configurations (N) searched at each iteration;

Algorithm 4 Tabu search

1: **Inputs:** S, f, C ▷ S is the initial solution, f is the evaluation function, C contains control parameters ($maxit$, L and N)
2: $maxit \leftarrow get\_maxit(C)$ ▷ maximum number of iterations
3: $L \leftarrow get\_L(C)$ ▷ length of the tabu list
4: $N \leftarrow get\_N(C)$ ▷ number of neighbor configurations to check at each iteration
5: $List \leftarrow \{S\}$ ▷ tabu list (first in, first-out queue)
6: $i \leftarrow 0$ ▷ i is the number of iterations of the method
7: **while** $i < maxit$ **do** ▷ $maxit$ is the termination criterion
8: $CList \leftarrow \{\}$ ▷ candidate list
9: **for** $j = 1 \rightarrow N$ **do** ▷ cycle j from 1 to N
10: $S' \leftarrow change(S, C)$ ▷ new solution
11: **if** $S' \notin List$ **then** $CList \leftarrow CList \cup S'$ ▷ add S' into $CList$
12: **end if**
13: **end for**
14: $S' \leftarrow best(CList, f)$ ▷ get best candidate solution
15: **if** $isbest(S', S, f)$ **then** ▷ if S' is better than S
16: $List \leftarrow List \cup S'$ ▷ enqueue S' into $List$
17: **if** $length(List) > L$ **then** $dequeue(L)$ ▷ remove oldest element
18: **end if**
19: $S \leftarrow S'$ ▷ set S as the best solution S'
20: **end if**
21: $i \leftarrow i + 1$
22: **end while**
23: **Output:** S ▷ the best solution

- **listSize**—length of the tabu list (L);
- **nRestarts**—maximum number of restarts in the intensification stage; and
- **repeatAll**—number of times to repeat the search.

The **tabuSearch** function returns a list with elements such as **$configKeep**—matrix with stored solutions, and **$eUtilityKeep**—vector with the respective evaluations.

To demonstrate this method, file **binary-tabu.R** optimizes the binary tasks of Sect. 1.8:

```
### binary-tabu.R file ###
library(tabuSearch) # load tabuSearch package

# tabu search for sum of bits:
sumbin=function(x) (sum(x)) # sum of bits
D=8 # dimension
s0=rep(0,D) # c(0,0,0,0,...)

cat("sum of bits (D=",D,")\n",sep="")
s1=tabuSearch(D,iters=2,objFunc=sumbin,config=s0,neigh=2,
              listSize=4,nRestarts=1)
b=which.max(s1$eUtilityKeep) # best index
cat("best:",s1$configKeep[b,],"f:",s1$eUtilityKeep[b],"\n")

# tabu search for max sin:
```

```
intbin=function(x) sum(2^(which(rev(x==1))-1))
maxsin=function(x) # max sin (explained in Chapter 3)
{ D=length(x);x=intbin(x); return(sin(pi*(as.numeric(x))/(2^D)))
  }
D=8
cat("max sin (D=",D,")\n",sep="")
s2=tabuSearch(D,iters=2,objFunc=maxsin,config=s0,neigh=2,
             listSize=4,nRestarts=1)
b=which.max(s2$eUtilityKeep) # best index
cat("best:",s2$configKeep[b,],"f:",s2$eUtilityKeep[b],"\n")
```

An example of file **binary-tabu.R** execution is:

```
> source("binary-tabu.R")
sum of bits (D=8)
best: 1 0 1 0 1 1 1 1 f: 6
max sin (D=8)
best: 0 1 1 0 1 1 1 0 f: 0.9757021
```

While few iterations were used, the method optimized solutions close to the optimum values (which are $f = 8$ for **sum of bits** and $f = 1$ for **max sin**).

The tabu search is also demonstrated for the **bag prices** integer task ($D = 5$). Given that **tabuSearch()** imposes some restrictions, adaptations are needed. The most relevant is the use of a binary representation, which in the demonstration is set to 10 digits per integer value (to cover the {\$1,\$2,...,\$1000} range). Also, since the associated search space includes infeasible solutions, a simple death-penalty scheme is used (Sect. 1.5), where $f = -\infty$ if any price is above \$1000. Finally, given that **tabuSearch()** does not include extra arguments to be passed to the evaluation function, these arguments (**D** and **Dim**) need to be explicitly defined before the tabu search method is executed. The adapted R code (file **bag-tabu.R**) is:

```
### bag-tabu.R file ###
library(tabuSearch) # load tabuSearch package
source("functions.R") # load the profit function

# tabu search for bag prices:
D=5 # dimension (number of prices)
MaxPrice=1000
Dim=ceiling(log(MaxPrice,2)) # size of each price (=10)
size=D*Dim # total number of bits (=50)
s0=sample(0:1,size,replace=TRUE) # initial search

intbin=function(x) # convert binary to integer
{ sum(2^(which(rev(x==1))-1)) } # explained in Chapter 3

bintbin=function(x) # convert binary to D prices
{ # note: D and Dim need to be set outside this function
  s=vector(length=D)
  for(i in 1:D) # convert x into s:
  { ini=(i-1)*Dim+1;end=ini+Dim-1
    s[i]=intbin(x[ini:end])
```

```
  }
  return(s)
}

bprofit=function(x) # profit for binary x
{ s=bintbin(x)
  if(sum(s>MaxPrice)>0) f=-Inf # death penalty
  else f=profit(s)
  return(f)
}

cat("initial:",bintbin(s0),"f:",bprofit(s0),"\n")
s=tabuSearch(size,iters=100,objFunc=bprofit,config=s0,neigh=4,
    listSize=16,nRestarts=1)
b=which.max(s$eUtilityKeep) # best index
cat("best:",bintbin(s$configKeep[b,]),"f:",s$eUtilityKeep[b],"\n"
    )
```

This code introduces the **ceiling()** R function that returns the closest upper integer. An execution example of file **bag-tabu.R** is:

```
> source("bag-tabu.R")
initial: 621 1005 880 884 435 f: -Inf
best: 419 428 442 425 382 f: 43050
```

In this case, the tabu search managed to improve an infeasible initial search point into a solution that is only 2% far from the optimum value ($f = 43899$).

4.5 Comparison of Local Search Methods

The comparison of optimization methods is not a trivial task. The *No Free Lunch (NFL)* theorem (Wolpert and Macready, 1997) states that all search methods have a similar global performance when compared over all possible functions. However, the set of all functions includes random and deceptive ones, which often are not interesting to be optimized (as discussed in Sect. 1.7). A constructive response to the theorem is to define a subset of "searchable" functions where the theorem does not hold, comparing the average performance of a several algorithms on this subset (Mendes, 2004). Yet, even if an interesting subset of functions and methods is selected, there are other relevant issues for a robust comparison: how to tune the control parameters of a method, such as the temperature T of the simulated annealing method (Sect. 4.6 addresses this issue) and which performance metrics and statistical tests should be adopted (discussed in this section).

Hence, rather than presenting a complete comparison, this section presents an R code example of how optimization methods can be compared, assuming some reasonable assumptions (if needed, these can be changed by the readers). The example uses one task, **rastrigin** benchmark with $D = 20$ (which is the most difficult real value task from Sect. 1.8), and it compares three methods: Monte Carlo (Sect. 3.4), hill climbing (Sect. 4.2), and simulated annealing (Sect. 4.3, using the

optim implementation). To avoid any bias toward a method, the same *change* function is used for hill climbing and simulated annealing strategies, and the default **optim** values ($T = 10$, $tmax = 10$) are adopted for the last search strategy. The same maximum number of iterations ($maxit = 10000$) is used for all methods. Rather than comparing just the final best value, the comparison is made throughout the search execution. Some measures of search execution can be deceptive, such as time elapsed, which might be dependent on the processor workload, or the number of iterations, whose computational effort depends on the type of search. Thus, the best value is stored for each evaluation function (from 1 to 10,000), as sequentially called by the method. Finally, a total of 50 runs are executed for each method, with the initial solutions randomly generated within the range $[-5.2, 5.2]$. To aggregate the results, the average and respective Student's t-test 95% confidence interval curves are computed for the best values. The comparison code (file **compare.R**) outputs a PDF result file (**comp-rastrigin.pdf**):

```
### compare.R file ###

source("hill.R") # get hchange
source("blind.R") # get fsearch
source("montecarlo.R") # get mcsearch
library(rminer) # get meanint

# comparison setup:
crastrigin=function(x) # x is a solution
{ f=10*length(x)+sum(x^2-10*cos(2*pi*x))
  # global variables: EV, BEST, F
  # global assignment code: <<-
  EV<<-EV+1 # increase evaluations
  if(f<BEST) BEST<<-f # update current BEST
  if(EV<=MAXIT) F[EV]<<-BEST # update BEST for EV
  return(f)
}

# experimental comparison setup:
Runs=50; D=20; MAXIT=10000
lower=rep(-5.2,D);upper=rep(5.2,D)
rchange1=function(par,lower,upper) # change for hclimbing
{ hchange(par,lower=lower,upper=upper,rnorm,
          mean=0,sd=0.5,round=FALSE) }
rchange2=function(par)             # change for optim
{ hchange(par,lower=lower,upper=upper,rnorm,
          mean=0,sd=0.5,round=FALSE) }
CHILL=list(maxit=MAXIT,REPORT=0) # control for hclimbing
CSANN=list(maxit=MAXIT,temp=10,trace=FALSE) # control for SANN
Methods=c("Monte Carlo","hill climbing","simulated annealing")

# run all optimizations and store results:
RES=vector("list",length(Methods)) # all results
for(m in 1:length(Methods))
   RES[[m]]=matrix(nrow=MAXIT,ncol=Runs)
```

```
for(R in 1:Runs) # cycle all runs
{ s=runif(D,-5.2,5.2) # initial search point
  EV=0; BEST=Inf; F=rep(NA,MAXIT) # reset these global vars.
  # Monte Carlo:
  mcsearch(fn=crastrigin,lower=lower,upper=upper,N=MAXIT)
  RES[[1]][,R]=F
  # hill climbing:
  EV=0; BEST=Inf; F=rep(NA,MAXIT)
  hclimbing(s,crastrigin,change=rchange1,lower=lower,
            upper=upper,control=CHILL,type="min")
  RES[[2]][,R]=F
  # SANN:
  EV=0; BEST=Inf; F=rep(NA,MAXIT)
  optim(s,crastrigin,method="SANN",gr=rchange2,control=CSANN)
  RES[[3]][,R]=F
}

# aggregate (average and confidence interval) results:
AV=matrix(nrow=MAXIT,ncol=length(Methods))
CI=AV
for(m in 1:length(Methods))
{
 for(i in 1:MAXIT)
 {
  mi=meanint(RES[[m]][i,]) # mean and confidence interval
  AV[i,m]=mi$mean;CI[i,m]=mi$int
 }
}

# show comparative PDF graph:

# plot a nice confidence interval bar:
#   x are the x-axis points
#   ylower and yupper are the lower and upper y-axis points
#   ... means other optional plot parameters (lty, etc.)
confbar=function(x,ylower,yupper,K=100,...)
{ segments(x-K,yupper,x+K,...)
  segments(x-K,ylower,x+K,...)
  segments(x,ylower,x,yupper,...)
}

pdf("comp-rastrigin.pdf",width=5,height=5)
par(mar=c(4.0,4.0,0.1,0.6)) # reduce default plot margin
MIN=min(AV-CI);MAX=max(AV+CI)
# 10.000 are too much points, thus two grids are used
# to improve clarity:
g1=seq(1,MAXIT,length.out=1000) # grid for lines
g2=seq(1,MAXIT,length.out=11) # grid for confbar
plot(g1,AV[g1,3],ylim=c(MIN,MAX),type="l",lwd=2,
     ylab="average best",xlab="number of evaluations")
confbar(g2,AV[g2,3]-CI[g2,3],AV[g2,3]+CI[g2,3])
lines(g1,AV[g1,2],lwd=2,lty=2)
confbar(g2,AV[g2,2]-CI[g2,2],AV[g2,2]+CI[g2,2])
```

```
lines(g1,AV[g1,1],lwd=2,lty=3)
confbar(g2,AV[g2,1]-CI[g2,1],AV[g2,1]+CI[g2,1])
legend("topright",legend=rev(Methods),lwd=2,lty=1:3)
dev.off() # close the PDF device
```

Given that some optimization functions (e.g., **optim**) are restrictive in terms of the parameters that can be used as inputs, the evaluation function is adapted to perform global assignments (operator **<<-**, Sect. 2.3) to the number of evaluations (**EV**), best value (**BEST**), and vector of best function values (**F**). The results are stored in a vector list of size 3, each element with a matrix $maxit \times runs$. Two similar *change* functions are defined, since **optim** does not allow the definition of additional arguments to be passed to **gr**. The code introduces some new R functions:

- **meanint** (from package **rminer**)—computes the mean and t-test confidence intervals;
- **segments**—draws a segment;
- **par**—sets graphical parameters used by **plot**; and
- **lines**—joints points into line segments.

The result execution of file **compare.R** is presented in Fig. 4.4. Initially, all three methods present a fast and similar convergence. However, after around 2000 evaluations, the hill climbing and simulated annealing methods start to outperform the random search (Monte Carlo). The confidence interval bars show that after around 4000 evaluations, the local search methods are statistically significantly better when compared with Monte Carlo. In this experiment, simulated annealing produces only a slight best average result, and the differences are not statistically significant when compared with hill climbing (since confidence intervals overlap).

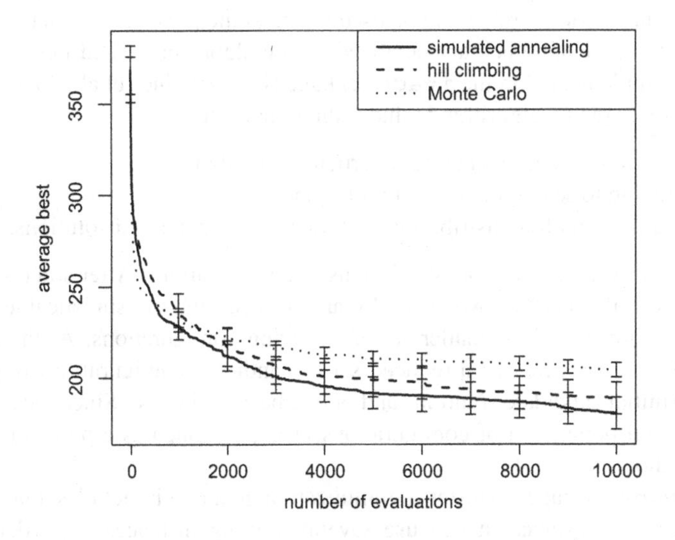

Fig. 4.4 Local search comparison example for the **rastrigin** task ($D = 20$)

4.6 Tuning Optimization Parameters

Most of the modern optimization methods presented in this book include a large number of parameters that need to be set a priori by the users and that affect the search performance. For example, while the stochastic hill climbing (Sect. 4.2) includes just two numeric parameters (P—probability and $maxit$—the maximum number of iterations), the simulation annealing implemented by the **NMOF** package contains up to seven numeric arguments:

- `nT`—number of temperatures (default is 10);
- `nI`—total number of iterations ($maxit$, default is `NULL`);
- `nD`—number of random steps to change the temperature (default is 2000);
- `initT`—initial temperature (default is `NULL`);
- `finalT`—final temperature (default is 0.0);
- `alpha`—cooling constant (default is 0.9); and
- `mStep`—step multiplier (default is 1.0).

Non-expert users typically use the default parameter values or perform some trial-and-error experiments. However, the default parametrization has some issues. For instance, there are algorithm implementations that do not include default values. Moreover, the suggested default values were often tuned for some specific tasks that can be quite different from the currently analyzed optimization goal, which can result in an inefficient search. Regarding the trial-and-error experiments, these are often conducted manually and thus tend to be ad hoc, based on the analyst's experience or intuition.

A better and replicable approach is to automate the parameter tuning procedure, which is a second order optimization task. In effect, any optimization method can be applied to tune the parameters of another optimization method. For instance, a simpler Monte Carlo search could be used to adjust the parameters of a tabu search. In this section, we approach a specialized tuning algorithm, called iterated racing, and that is implemented in the **irace** package (López-Ibáñez et al., 2016).

The iterated racing algorithm includes three main steps:

1. sampling of new solutions using a particular distribution;
2. use of racing to select the best solutions; and
3. updating the sampling distribution such that it biases the best solutions.

A race starts with a population of solutions or configurations. After some steps, the configurations that perform worst are discarded, according to a statistical test. Thus, the race continues with a smaller set of surviving configurations. At the end, the algorithm converges toward a reduced set of candidate configurations (e.g., there can be 10 initial candidate solutions and only 2 best final ones). Since the algorithm uses an initial population of configurations, it is positioned as a population based search method in Fig. 1.4.

The **irace** package performs a minimization goal and it includes a large range of features. For instance, it can use several training instances (e.g., datasets in different files) and work with any executable program, provided that appropriate execution scripts are defined (e.g., using Unix shell scripting). This section demonstrates a simpler **irace** usage with R code and no instances. The implemented **rastrigin-irace.R** code uses the **irace** method to tune four simulated annealing parameters of the **NMOF** package (**nT**, **nD**, **initT**, and **alpha**) when optimizing the **rastrigin** task:

```
### rastrigin-irace.R file ###

# tune SAopt parameter for rastrigin D=30

library(NMOF) # load NMOF
library(irace) # load irace
source("hill.R") # get hchange

# real value rastrigin function with a dimension of 30:
rastrigin=function(x) 10*length(x)+sum(x^2-10*cos(2*pi*x))
D=30

set.seed(123) # set for replicability

# setup rastrigin experiments:
lower=rep(-5.2,D);upper=rep(5.2,D) # lower and upper bounds
par=runif(D,lower[1],upper[1]) # fixed initial search point

rchange=function(par) # real change
{ D=length(par)
  # upper and lower are defined globally
  hchange(par,lower=lower,upper=upper,rnorm,mean=0,sd=1)
}

# irace elements: parameters and scenario --------------------
# 4 parameters for irace: i - integer, r - real value
parameters.txt='
nT "" i (1, 20)
nD "" i (1, 4000)
initT "" r (0.1, 1000.0)
alpha "" r (0.0, 1.0)
'
parameters=readParameters(text=parameters.txt)

# targetRunner function: call SAopt for an irace configuration
rastrigin_run=function(experiment,scenario)
{
  C=experiment$configuration # get current irace configuration
  # set algo with the irace configuration values
  algo=list(x0=par,nI=100,neighbour=rchange,      # fixed part
            printDetail=FALSE,printBar=FALSE,     # fixed part
            nT=C$nT,nD=C$nD,initT=C$initT,alpha=C$alpha) # irace
  res=SAopt(rastrigin,algo) # call SAopt
  return(list(cost=res$OFvalue)) # output list required by irace
}

scenario=list(targetRunner=rastrigin_run,
              instances=1, # not used but needs to be defined
```

```
              maxExperiments=200, # 200 calls to targetRunner
              logFile = "") # do not create log file
# ----------------------------------------------------------

# run SAopt with default values:
cat("default SAopt:\n")
algo1=list(x0=par,nI=100,neighbour=rchange,printDetail=FALSE,
    printBar=FALSE)
res=SAopt(rastrigin,algo1) # call SAopt
cat(" evaluation value:",res$OFvalue,"\n")

# run irace:
cat("irace SAopt:\n")
s=irace(scenario=scenario,parameters=parameters)
# show best configurations:
configurations.print(s)
# get best
b=removeConfigurationsMetaData(s[1,])
print(b)
algo2=list(x0=par,nI=100,neighbour=rchange,
           printDetail=FALSE,printBar=FALSE,
           nT=b$nT,initT=b$initT,alpha=b$alpha)
res=SAopt(rastrigin,algo2) # call SAopt
cat(" evaluation value:",res$OFvalue,"\n")
```

The **rastrigin** task is set similarly to what was defined in Sect. 4.3 (e.g., with a dimension of D =30). The **rastrigin-irace.R** executes two main simulated annealing searches: using the default **SAopt** values and using the values tuned by the **irace** tool. The **irace** main function requires two arguments: **scenario** and **parameters**. The latter argument defines the search space for the parameters. The demonstration code searches for two integers (**nT** $\in \{1, 2, \ldots, 20\}$ and **nD** $\in \{1, 2, \ldots, 2000\}$) and two real value arguments (**initT** $\in [0.1, 1000.0]$ and **alpha** $\in [0.0, 1.0]$). It should be noted that the **NMOF** package help does not provide explicit lower and upper values for these arguments; thus, a relatively large search space was used in the demonstration. The **readParameters()** function creates a parameter list from a file or character string that adopts the **irace** parameter syntax (e.g., one line per parameter). In this case, it uses the previously defined **parameters.txt** string. As for the former main argument (**scenario**), it consists of a list with several components, including:

- **$targetRunner**—a function that evaluates a target configuration, needing to return a list with a **$cost** component (the evaluation value, to be minimized); and
- **$maxExperiments**—maximum number of experiments, used to terminate the **irace** execution.

The implemented **rastrigin_run()** runner only uses the **experiment** argument, which contains the simulated annealing configuration (for **SAopt**). The line **C=experiment$configuration** defines an internal object that is a **data.frame** with one row and that can be manipulated as a list (e.g., **C$nT** gets access to the **nT** value of the current **irace** tested configuration). To speed up the demonstration, the code uses a relatively small number of iterations (**nI=100**) and **irace** experiments (**maxExperiments=200**). The other **irace** functions used in the code are:

- **configurations.print()**—prints the best **irace** configurations; and
- **removeConfigurationsMetaData()**—simplifies the **irace** configuration (e.g., for print purposes).

The **rastrigin-irace.R** execution produces a large console text. Thus, to simplify the analysis, only a reduced portion of the execution output text is shown:

```
> source("rastrigin-irace.R")
default SAopt:
 evaluation value: 406.1372
irace SAopt:
# 2020-06-02 11:34:44 WEST: Initialization
# Elitist race
# Elitist new instances: 1
# Elitist limit: 2
# nbIterations: 4
# minNbSurvival: 4
# nbParameters: 4
# seed: 1516206640
# confidence level: 0.95
# budget: 200
# mu: 5
# deterministic: FALSE

# 2020-06-02 11:34:44 WEST: Iteration 1 of 4
# experimentsUsedSoFar: 0
# remainingBudget: 200
# currentBudget: 50
# nbConfigurations: 8
# ...
# ... (skipped irace output text)
# ...
# Iteration: 5
# nbIterations: 5
# experimentsUsedSoFar: 193
# timeUsed: 0
# remainingBudget: 7
# currentBudget: 7
# number of elites: 4
# nbConfigurations: 4
   nT   nD    initT  alpha
23 14  300 177.5743 0.2077
2  19 3458 882.3748 0.5469
5  16 1194 465.4429 0.5134
22 17 3626 998.9799 0.4729
   nT  nD    initT  alpha
23 14 300 177.5743 0.2077
 evaluation value: 314.2668
```

The **irace** execution uses an initial value of 8 configurations in its first iteration and ends with 4 configurations in its fourth iteration, after testing 193 experiments (**targetRunner** executions). The default **SAopt** values resulted in an optimized **rastrigin** function of 406.1, while the **irace** tuning (**nT=14**, **nD=300**, **initT=177.5743**, and **alpha=0.2077**) minimized a smaller **rastrigin** value (314.3). It is interesting to note that while only 100 iterations were executed, the tuned simulated annealing is capable of optimizing a value that is smaller when compared with the result achieved when running **SAopt** with default parameters and a much larger number of iterations (Sect. 4.3, 10000 iterations and best evaluation of 340.1).

4.7 Command Summary

GenSA	package with generalized simulated annealing
GenSA()	generalized simulated annealing (package **GenSA**)
NMOF	package with numerical methods and optimization in finance
SAopt()	simulated annealing function (package **NMOF**)
ceiling()	returns the closest upper integer
configurations.print()	prints best irace configurations (package **irace**)
hchange()	slight random change of a vector (chapter file **"hill.R"**)
hclimbing()	standard hill climbing search (chapter file **"hill.R"**)
ifelse()	conditional element selection
irace	package with iterated racing
irace()	iterated racing function (package **irace**)
lines()	joints points into line segments
meanint()	computes the mean and Student's t-test confidence intervals (package **rminer**)
optim()	general-purpose optimization (includes simulated annealing)
par()	sets or queries graphical parameters used by **plot()**
readParameters()	reads parameters from a string (package **irace**)
removeConfigurationsMetaData()	simplifies an irace configuration (package **irace**)
rminer	package for simpler use of data mining (classification and regression) methods
sa_hclimbing()	the steepest ascent hill climbing (chapter file **"hill2.R"**)
segments()	draws a segment line
st_hclimbing()	stochastic hill climbing (chapter file **"hill2.R"**)
stats	R base package that includes **optim()**
tabuSearch	package for tabu search
tabuSearch()	tabu search for binary string maximization (package **tabuSearch**)

4.8 Exercises

4.1 Using file **hill2.R**, create the new function **s2_hclimbing** that merges both the steepest ascent and stochastic hill climbing algorithms. Thus, the control parameter should define both **control$N** and **control$P**. The function should implement N random searches within the neighborhood of **par**, selecting the best neighbor. Then, it should accept the neighbor with a probability P or the best between the neighbor and current solutions.

4.2 Explore the optimization of the binary **max sin** task with a higher dimension ($D = 16$), under hill climbing, simulated annealing, and tabu search methods. Use the zero vector as the starting point and a maximum of 20 iterations. Show the optimized solutions and evaluation values.

4.3 Execute the optimization of the **rastrigin** function ($D = 8$) with the **tabuSearch** function. Adopt a binary representation such that each dimension value is encoded into 8 bits, denoting any of the 256 regular levels within the range $[-5.2, 5.2]$. Use the control parameters: $maxit = 500$, $N = 8$, $L = 8$, and **nRestarts=1** and a randomly generated initial point.

4.4 Use the iterated racing algorithm to tune the **neigh** and **listSize** parameters of the **tabuSearch()** function for **bag prices**. Suggestion: adapt the **bag-tabu.R** demonstration file from Sect. 4.4.

Chapter 5
Population Based Search

5.1 Introduction

In previous chapter, several local based search methods were presented, such as hill climbing, simulated annealing, and tabu search. All these methods are single-state, thus operating their effort around the neighborhood of a current solution. This type of search is simple and quite often efficient (Michalewicz et al., 2006). However, there is another interesting class of search methods, known as population based search, that uses a pool of candidate solutions rather than a single search point. Thus, population based methods tend to require more computation when compared with simpler local methods, although they tend to work better as global optimization methods, quickly finding interesting regions of the search space (Michalewicz and Fogel, 2004).

As shown in Fig. 5.1, population based methods tend to explore more distinct regions of the search space, when compared with single-state methods. As consequence, more diversity can be reached in terms of setting new solutions, which can be created not only by slightly changing each individual search point but also by combining attributes related with two (or more) search points.

The main difference between population based methods is set in terms of: how solutions are represented and what attributes are stored for each search point; how new solutions are created; and how the best population individuals are selected. Most population based methods are naturally inspired (Luke, 2015). Natural phenomena such as genetics, natural selection, and collective behavior of animals has led to optimization techniques such as genetic and evolutionary algorithms, genetic programming, estimation of distribution, differential evolution, and particle swarm optimization. This chapter describes all these methods and examples of their applications using the R tool.

P. Cortez, *Modern Optimization with R*, Use R!,
https://doi.org/10.1007/978-3-030-72819-9_5

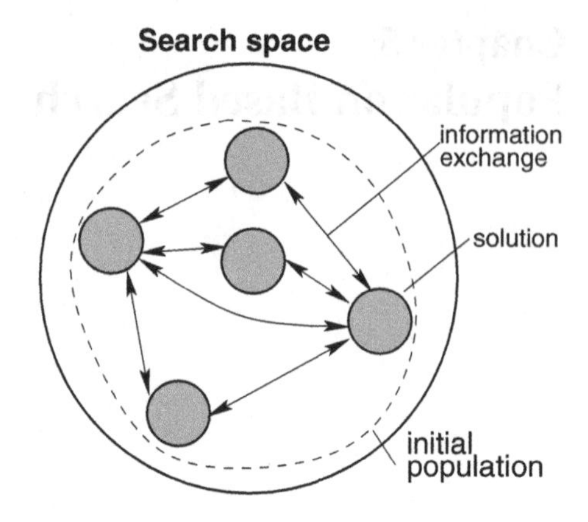

Fig. 5.1 Example of a population based search strategy

5.2 Genetic and Evolutionary Algorithms

Evolutionary computation denotes several optimization algorithms inspired in the natural selection phenomenon and that include a population of competing solutions. It is this "survival of the fittest" mechanism that distinguishes these algorithms from random search, allowing to quickly reach quality solutions within the search space (Michalewicz, 1996). Although it is not always clearly defined, the distinction among evolutionary computation methods is mostly based on how to represent a solution and how new solutions are created. Genetic algorithms were proposed by J. Holland (1975). The original method worked only on binary representations and adopted massively the crossover operator for generating new solutions. More recently, the term evolutionary algorithm was adopted to address genetic algorithm variants that include real value representations and that adopt flexible genetic operators, ranging from heavy use of crossover to only mutation changes (Michalewicz, 1996).

There is a biological terminology associated with evolutionary computation methods (Luke, 2015). Typically, the generic structure of Algorithm 1 is followed. First, an initial *population* is defined (S), which contains a pool of *individuals* (the candidate solutions). Often, the initial population is set randomly, although domain knowledge can also be used to set the initial solutions (as shown in Sect. 7.2). Then, there is a main evolutionary cycle, in which new *offspring* is created and best individuals are selected for the next iteration. Most algorithms adopt the generational approach, where in each iteration, called *generation*, all or a portion of the population individuals are replaced by offspring solutions. Steady-state algorithms create just one offspring solution in each iteration, which replaces the worst fit individual.

The *genotype*, *genome*, or *chromosome* denotes the individual data structure representation. A *gene* is a value position in such representation and an *allele* is a particular value for a gene (e.g., binary, integer, or real value). The evaluation function, also known as *fitness*, allows to measure the quality of an individual and the *phenotype* represents how the individual operates during fitness assessment. The creation of new solutions (*offspring*) is called *breeding* and occurs due to the application of *genetic* operators, such as *crossover* and *mutation*. Crossover involves selecting two or more *parent* solutions in order to generate *children*, while mutation often performs a slight change to a single individual. The *selection* operation sets how individuals are selected for *breeding* and for *survival* (going into the next *generation*).

The customization of the evolutionary search can involve setting the internal parameters of an algorithm (e.g., population size), selecting the breeding and selection operators, or even defining how the evolutionary main cycle works. In particular, the breeding operators can substantially improve the search result. Examples of generic operators that are detailed in this section are: crossover—one-point and uniform crossover; and mutation—uniform random and Gaussian mutation. Some operators are exclusive of some types of representations, such as the permutation based representations (discussed in Sect. 7.2). Moreover, there are evolutionary methods that mostly adopt mutation to create the offspring, such as evolution strategies (Luke, 2015). In contrast, there is a top performing Traveling Salesman Problem (TSP) evolutionary approach that only uses a specialized crossover (and no mutation) (Nagata and Kobayashi, 2013). Furthermore, often breeding operators should produce slight and not abrupt changes to the selected individuals. For instance, in some applications, uniform random mutation or crossover can be too destructive, erasing the informational value of the selected individual/gene and working more as a random search. As for selection operators, some examples include the roulette wheel, ranking, and tournament (Goldberg and Deb, 1990). The classical roulette wheel assumes a probabilistic selection method, where individuals are sampled in proportion to their fitness values. The linear rank selection works by first ranking the individuals according to their fitness and then selecting the individuals in proportion to their ranks. The tournament selection works by randomly sampling k individuals (solutions) from the population and then selecting the best n individuals. In recent years, the tournament selection has become a popular selection option for evolutionary computation methods due to three main reasons (Luke, 2015): it is less sensitive to particular values of the fitness function; it is simple and works well with parallel algorithms; and it can be tuned by changing the k value (the $k = 2$ is often used with genetic algorithms).

There are different R packages that implement genetic and evolutionary algorithms. This chapter details initially one of the first R implementations, which dates back to 2005 and is provided by the **genalg** package (Lucasius and Kateman, 1993). The package handles minimization tasks and contains two relevant functions: **rbga.bin**, for binary chromosomes; and **rbga**, for real value representations. The **genalg** main pseudo-code for **rbga.bin()** and **rbga()** is presented in Algorithm 5 and is detailed in the next paragraphs.

The *initialization* function creates a random population of N_P individuals (argument **popSize**), under a particular distribution (uniform or other). Each individual contains a fixed length chromosome (with L_S genes), defined by **size** for **rbga.bin** or by the length of the lower bound values (**stringMin**)

Algorithm 5 Genetic/evolutionary algorithm as implemented by the **genalg** package

1: **Inputs:** f, C ▷ f is the evaluation (fitness) function, C includes control parameters
2: $P \leftarrow initialization(C)$ ▷ random initial population
3: $N_P \leftarrow get\_population\_size(C)$ ▷ population size
4: $E \leftarrow get\_elitism(C)$ ▷ number of best individuals kept (elitism)
5: $i \leftarrow 0$ ▷ i is the number of iterations of the method
6: **while** $i < maxit$ **do**
7: $F_P \leftarrow f(P)$ ▷ evaluate current population
8: $P_E \leftarrow best(P, F_P, E)$ ▷ set the elitism population (lowest E fitness values)
9: $Parents \leftarrow selectparents(P, F_P, N_P - E)$ ▷ select $N_P - E$ parents from current population
10: $Children \leftarrow crossover(Parents, C)$ ▷ create $N_P - E$ children solutions
11: $Children \leftarrow mutation(Children, maxit, i)$ ▷ apply the mutation operator to the children
12: $P \leftarrow E \cup Children$ ▷ set the next population
13: $i \leftarrow i + 1$
14: **end while**
15: **Output:** P ▷ last population

for **rbga**. For **rbga**, the initial values are randomly generated within the lower (**stringMin**) and upper (**stringMax**) bounds. The optional argument **suggestions** can be used to include a priori knowledge, by defining an explicit initial matrix with up to N_P solutions. The function **rbga.bin()** includes also the argument **zeroToOneRatio** that denotes the probability for choosing a zero for mutations and population initialization. The ideal number of individuals (N_P, argument **popSize**) is problem dependent. Often, this value is set taking into account the chromosome length (L_S), computational effort, and preliminary experiments. In practice, common population size values are $N_P \in \{20, 50, 100, 200, 500, 1000, 2000\}$.

The algorithm runs for $maxit$ generations (argument **iters**). The *selection for survival* procedure is based on an *elitism* scheme and the created offspring (*Children*). If an *elitism* scheme is adopted ($E > 0$, value set by argument **elitism**), then the best E individuals from the current population always pass to the next generation (the default is **elitism**=20% of the population size). The remaining next population individuals are created by applying a crossover and then a mutation. The pseudo *selectparents* function aims at the selection of $N_P - E$ parents from the current population, in order to apply the crossover operator. The **rbga** algorithm performs a uniform random choice of parents, while **rbga.bin** executes a probabilistic selection of the fittest individuals. This selection works by ranking first the current population (according to the fitness values) and then performing a random selection of parents according to a density normal distribution. The respective R code for generating such probabilities is **dnorm(1:popSize, mean = 0, sd = (popSize/3))**. For instance, if $N_P = 4$ then the probabilities for choosing parents from the ranked population are

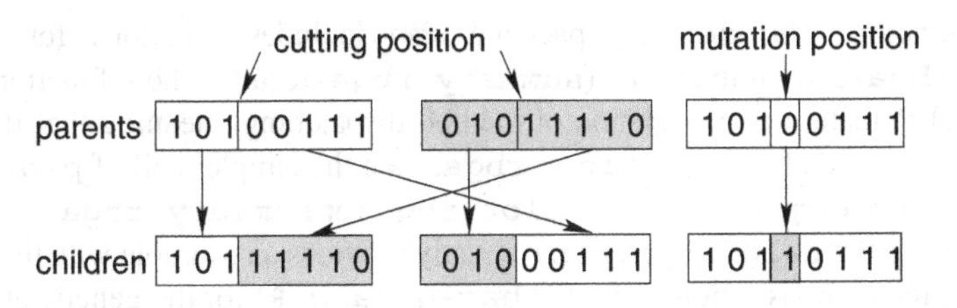

Fig. 5.2 Example of binary one-point crossover (left) and bit mutation (right) operators

$(0.23, 0.1, 0.02, 0.0)$, where 0.23 denotes the probability for the best individual. It should be noted that this probabilistic selection is a nonlinear rank method.

To create new individuals, both **rbga** and **rbga.bin** adopt the same one-point crossover, which was originally proposed in Holland (1975). The original one-point operator works by first selecting two parents and a random cutting position and then creating two children, each with distinct portions of the parents, as exemplified in the left of Fig. 5.2. As a specific **genalg** package implementation, only the first one-point crossover child is inserted into the new population and thus $N_P - E$ crossover operations are executed (Sect. 7.2 shows an implementation that inserts the two children generated from a crossover). Next, a mutation operator is executed over the children, where each gene can be changed with a small probability (set by **mutationChange**). By default, **mutationChange** is set to **1/(size+1)**. Once a gene is mutated, the new value is set differently according to the chromosome representation type. In the binary version, the new bit value is randomly set taking into account the **zeroToOneRatio** value. The right of Fig. 5.2 shows an example of a binary bit mutation. Mutated real values are obtained by first computing $g' = 0.67 \times r_d \times d_f \times R_g$, where $r_d \in \{-1, 1\}$ is a random direction, $d_f = (maxit - i)/maxit$ is a dampening factor, and $R_g = \max(g) - \min(g)$ is the range of the gene (e.g., $\max(g)$ denotes the upper bound for g, as set by **StringMax**). If g' lies outside the lower (**StringMin**) or upper bounds (**StringMax**), then $g' = \min(g) + \mathcal{U}(0, 1) \times R_g$. More details can checked by accessing the **rbga** package source code (`> getAnywhere(rbga.bin)` and `> getAnywhere(rbga)`).

The **rbga.bin** and **rba** functions include four additional parameters:

- **monitorFunc**—monitoring function (e.g., could be used to compute useful statistics, such as population diversity measures), applied after each generation;
- **evalFunc**—evaluation (or fitness) function;
- **showSettings**—if **TRUE** then the genetic algorithm parameters are shown (e.g., N_P); and
- **verbose**—if **TRUE** then more text about the search evolution is displayed.

The result of executing **rbga.bin** and **rba** is a list with components such as: **$population**—last population; **$evaluations**—last fitness values; **$best**—best value per generation; and **$mean**—mean fitness value per generation. The **genalg** package also includes functions for plotting (**plot.rbga**) and summarizing (**summary.rbga**) results. These functions adopt the useful R feature of S3 scheme of method dispatching, meaning that if **obj** is the object returned by **rbga.bin** or **rbga**, then the simpler call of **plot(obj)** (or **summary(obj)**) will execute **plot.rbga** (or **summary.rbga**).

Given that the **help(rbga.bin)** already provides an example with the **sum of bits** task, the demonstration code (file **bag-genalg.R**) for the genetic algorithm explores the **bag prices** ($D = 5$) problem of Sect. 1.8 (the code was adapted from the example given for the tabu search in Sect. 4.5):

```
### bag-genalg.R file ###
library(genalg) # load genalg package
source("functions.R") # load the profit function

# genetic algorithm search for bag prices:
D=5 # dimension (number of prices)
MaxPrice=1000
Dim=ceiling(log(MaxPrice,2)) # size of each price (=10)
size=D*Dim # total number of bits (=50)
intbin=function(x) # convert binary to integer
{ sum(2^(which(rev(x==1))-1)) } # explained in Chapter 3
bintbin=function(x) # convert binary to D prices
{ # note: D and Dim need to be set outside this function
  s=vector(length=D)
  for(i in 1:D) # convert x into s:
  { ini=(i-1)*Dim+1;end=ini+Dim-1
    s[i]=intbin(x[ini:end])
  }
  return(s)
}
bprofit=function(x) # profit for binary x
{ s=bintbin(x)
  s=ifelse(s>MaxPrice,MaxPrice,s) # repair!
  f=-profit(s) # minimization task!
  return(f)
}
# genetic algorithm execution:
G=rbga.bin(size=size,popSize=50,iters=100,zeroToOneRatio=1,
    evalFunc=bprofit,elitism=1)
# show results:
b=which.min(G$evaluations) # best individual
cat("best:",bintbin(G$population[b,]),"f:",-G$evaluations[b],
    "\n")
pdf("genalg1.pdf") # personalized plot of G results
plot(-G$best,type="l",lwd=2,ylab="profit",xlab="generations")
lines(-G$mean,lty=2,lwd=2)
legend("bottomright",c("best","mean"),lty=1:2,lwd=2)
dev.off()
summary(G,echo=TRUE) # same as summary.rbga
```

Similarly to the tabu search example, 10 binary digits are used to encode each price. The evaluation function (**bprofit**) was adapted with two changes. First, a repair strategy was adopted for handling infeasible prices, where high prices are limited into the **MaxPrice** upper bound. Second, the **profit** function is multiplied by −1, since **genalg** only handles minimization tasks. The last code lines show results in terms of the best solution and summary of the genetic algorithm execution. Also, the code creates a plot showing the evolution of the best and mean profit values. An example of executing file **bag-genalg.R** is:

```
> source("bag-genalg.R")
best: 427 404 391 391 395 f: 43830
GA Settings
  Type                   = binary chromosome
  Population size        = 50
  Number of Generations  = 100
  Elitism                = 1
  Mutation Chance        = 0.0196078431372549

Search Domain
  Var 1 = [,]
  Var 0 = [,]

GA Results
  Best Solution : 0 1 1 0 1 0 1 0 1 1 0 1 1 0 0 1 0 1 0 0 0 1 1 0
       0 0 0 1 1 1 0 1 1 0 0 0 0 1 1 1 0 1 1 0 0 0 1 0 1 1
```

Using 100 generations, the genetic algorithm improved the initial population (randomly set) best profit from \$39646 (**-1*G$best[1]**) to \$43830, with the best fitness value being very close to the optimum (profit of \$43899). Figure 5.3 shows the best and mean profit values during the 100 generations.

The **rbga** demonstration code is related with the **sphere** ($D = 2$) task (file **sphere-genalg.R**):

```
### sphere-genalg.R file ###
library(genalg) # load genalg

# evolutionary algorithm for sphere:
sphere=function(x) sum(x^2)
D=2
monitor=function(obj)
{ if(i==1)
    { plot(obj$population,xlim=c(-5.2,5.2),ylim=c(-5.2,5.2),
           xlab="x1",ylab="x2",type="p",pch=16,
           col=gray(1-i/maxit))
    }
  else if(i%%K==0) # add points to the plot
     points(obj$population,pch=16,col=gray(1-i/maxit))
  i<<-i+1 # global update
}

maxit=100
K=5 # show population values every K generations
i=1 # initial generation
```

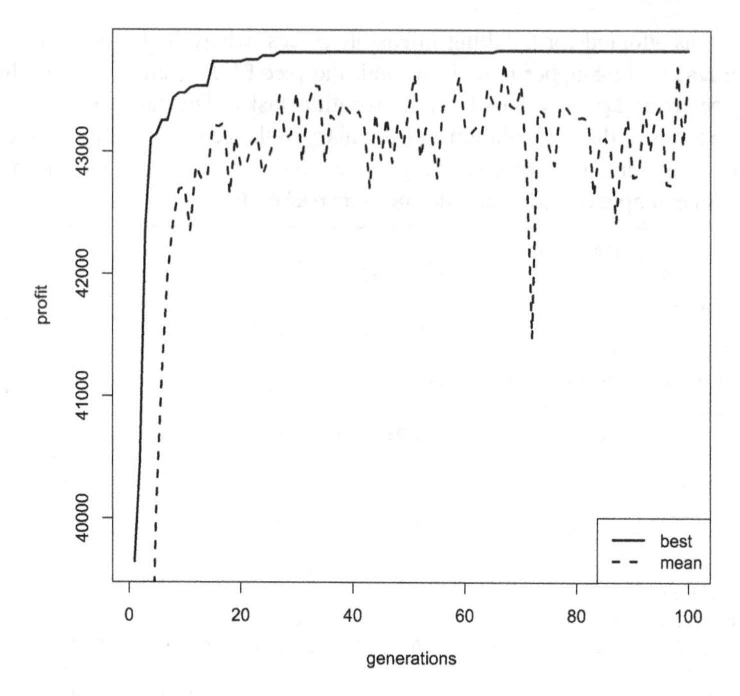

Fig. 5.3 Example of evolution of a genetic algorithm for task **bag prices**

```
# evolutionary algorithm execution:
pdf("genalg2.pdf",width=5,height=5)
set.seed(12345) # set for replicability
E=rbga(rep(-5.2,D),rep(5.2,D),popSize=5,iters=maxit,
       monitorFunc=monitor,evalFunc=sphere)
b=which.min(E$evaluations) # best individual
cat("best:",E$population[b,],"f:",E$evaluations[b],"\n")
dev.off()
```

In this example, the **monitor** argument is used to plot the population of solutions every **K** generations, using a coloring scheme that ranges from light gray (initial population) to dark (last generation). This gradient coloring is achieved using the **gray()** R function, which creates gray colors between 1 (white) and 0 (black). In the first call to **monitor**, the **plot()** function sets the initial graph. Then, in subsequent calls, when global object **i** is a multiple of **K**, additional points (representing current population values) are added to the graph by using the **points** R function. The use of the **set.seed** command (setting the R random seed) is adopted here only for reproducibility purposes, i.e. readers that execute this code should get the same results. The execution of file **sphere-genalg.R** is:

```
> source("sphere-genalg.R")
best: 0.02438428 0.09593309 f: 0.009797751
```

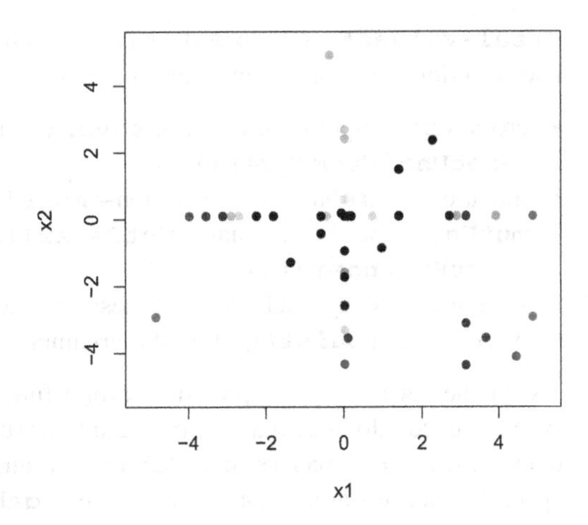

Fig. 5.4 Example of an evolutionary algorithm search for **sphere** ($D = 2$)

Although a very small population is used ($N_P = 5$, minimum value accepted by **rbga**), the evolved solution of $s = (0.024, 0.096)$ and $f = 0.0098$ is very close to the optimum ($f = 0$). Figure 5.4 presents the result of the plot, showing that darker points converge toward the optimum point (origin).

Given the popularity of evolutionary computation, there are several other R packages that implement genetic and evolutionary algorithms, such as:

- **ecr**—for single and multi-objective optimization with binary, real value, and permutation representations (sequential and fitness based parallel execution) (Bossek, 2017);
- **GA**—flexible genetic algorithms with binary, real value, and permutation representations (sequential and several parallel execution modes, including the island model) (Scrucca, 2017);
- **mcga**—it uses a byte representation of variables for a fast real value optimization (Satman, 2013); and
- **NMOF**—includes a simple genetic algorithm (binary representation) for minimizing functions (sequential and fitness based parallel execution) (Gilli et al., 2019).

An updated list of packages that implement evolutionary computation and other modern optimization methods can be found at the CRAN task view for optimization (https://cran.r-project.org/web/views/Optimization.html).

When compared with other evolutionary computation implementations, the **ecr** and **GA** packages are more flexible, providing distinct representation types, breeding operators, and selection methods. This is important because in complex real-world tasks the performance of evolutionary computation methods can depend much on a proper customization of the evolutionary search. The **ecr** implementation includes the **representation** argument of its **ecr()** function that allows to adopt **"binary"**, **"float"**, **"permutation"**, and even a **"custom"** type (defined by the user). Similarly, the **GA** package allows to work with three representation types by setting the **type** argument of its main **ga()** function to: **"binary"**,

"real-valued", or **"permutation"**. Both packages provide several breeding and selection operators. Some **ecr** operators are:

- crossover: one-point—**recCrossover()**, and uniform crossover —**recUnifCrossover()**;
- mutation: random bit mutation—**mutBitflip()**, random uniform—**mutUniform()**, Gaussian—**mutGauss()**, and polynomial —**mutPolynomial()**;
- selection: greedy (selection of best individuals)—**selGreedy()**, roulette wheel—**selRoulette()**, and tournament—**selTournament()**.

As for the **GA** package, it provides several functions, one for each representation type, grouped into three main help descriptions (use **help**(*description*) to get their details): **ga_Crossover**, **ga_Mutation**, and **ga_Selection**. Examples of specific functions are: one-point crossover—**gabin_spCrossover()**, random uniform mutation—**gareal_raMutation()**, and linear rank selection—**gaperm_lrSelection()**. In addition, the **ecr** package includes several auxiliary functions (e.g., **evaluateFitness** and **generateOffspring**) that can be used to create a customized evolutionary cycle, such as exemplified in Sect. 7.2.

In this section, all demonstration examples assume only a sequential execution (using a single processor) of the algorithms, since Sect. 5.9 is specifically devoted to parallel execution examples. The distinct genetic and evolutionary algorithm R packages are first explored with the **max sin** binary representation task (Sect. 1.8), with a dimension of $D = 20$ bits. The developed R code (file **maxsin-ga.R**) attempts to execute the same evolution procedure with the different package implementations. However, while these implementations share some evolution properties, they also contain substantial differences. For instance, all genetic algorithm packages are generational but there are some differences in the way the offspring replaces the current population individuals. The **genalg** package applies the crossover to all population solutions except the elitism individuals (E). In **ecr** and **GA**, it is possible to define a crossover probability, which defines the crossover chance between two pairs of individuals. Moreover, **genalg** only uses the one-point crossover, while the other packages allow other crossover possibilities, such as the uniform crossover. This crossover is adopted in the **rastrigin** example code and it selects a random number of bits (defining a binary mask) that are passed from the parents to the offspring (as shown in Fig. 5.5). Also, **genalg** and **NMOF** packages perform a bit mutation probability (which should be a small value), while **ecr** and **GA** use instead an individual mutation probability. Both **ecr** and **GA** allow a flexible definition of the selection operator, while in **NMOF** the selection is fixed to be pairwise, where the best solution is kept between the parent and its offspring. To obtain a selection procedure more similar to what is executed by **genalg**, in this demonstration a nonlinear rank selection was adopted for the **GA** package, while **ecr** was configured with a roulette wheel selection. Furthermore, the **GA** package executes a maximization goal, **rbga** and **NMOF** perform a minimization goal and **ecr** allows both optimization goals.

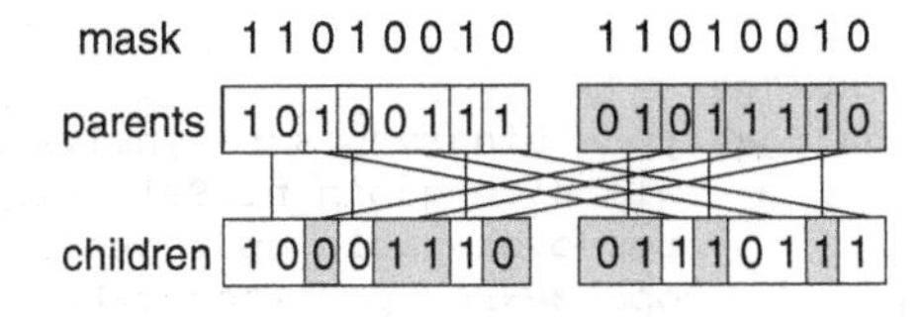

Fig. 5.5 Example of a binary uniform crossover operator

The **max sin** demonstration code is:

```
### maxsin-ga.R ###

library(genalg) # load rba.bin
library(ecr)    # load ecr
library(GA)     # load GA
library(NMOF)   # load GAopt

# maxsin task:
intbin=function(x) sum(2^(which(rev(x==1))-1))
maxsin=function(x) sinpi(intbin(x)/(2^D)) # maximization goal
maxsin2=function(x) -maxsin(x)            # minimization goal
D=20 # number of dimensions

# GA parameter setup:
Pop=20 # population size
Gen=100 # maximum number of generations
Eli=1L # elitism of one individual
PCx=1.0 # crossover probability for a solution
PMut1=0.01 # single bit mutation probability
PMut2=0.10  # mutation probability for a solution

# rbga.bin (minimization):
cat("rbga.bin:\n")
ga1=rbga.bin(size=D,popSize=Pop,iters=Gen,
            mutationChance=PMut1,zeroToOneRatio=1,
            evalFunc=maxsin2,elitism=Eli)
b=which.min(ga1$evaluations) # best individual
# show best solution and evaluation value:
cat("best:",ga1$population[b,],"f:",-ga1$evaluations[b],"\n")

# ecr (maximization):
cat("ecr:\n")
ga2=ecr(maxsin,minimize=FALSE,n.objectives=1,
        n.dim=D,n.bits=D,representation="binary",
        mu=Pop, # population size
        lambda=(Pop-Eli), # new solutions in each generation
        p.recomb=PCx,p.mut=PMut2,
        n.elite=Eli,
        parent.selector=selRoulette, # roulette wheel
        survival.selector=selRoulette, # roulette wheel
        recombinator=recCrossover, # one-point crossover
        terminators = list(stopOnIters(Gen)))
cat("best:",ga2$best.x[[1]],"f:",ga2$best.y,"\n")
```

```
# ga (maximization):
cat("ga:\n")
ga3=ga(type="binary",maxsin,nBits=D,
       selection=gabin_nlrSelection, # nonlinear rank
       crossover=gabin_spCrossover, # one-point crossover
       popSize=Pop,pcrossover=PCx,pmutation=PMut2,
       elitism=Eli,maxiter=Gen,monitor=FALSE)
cat("best:",ga3@solution,"f:",ga3@fitnessValue,"\n")

# GAopt (minimization):
cat("GAopt:\n")
algo=list(nB=D,nP=Pop,nG=Gen,
          crossover="onePoint", # one-point crossover
          prob=PMut1,printDetail=FALSE,printBar=FALSE)
ga4=GAopt(maxsin2,algo=algo)
cat("best:",as.numeric(ga4$xbest),"f:",-ga4$OFvalue,"\n")
```

The code loads first the distinct packages and then defines the **max sin** task, under two variants: **maxsin()**—standard maximization goal; and **maxsin2()**—for minimization algorithm implementations. In this demonstration, the **sinpi()** R function is used, which computes $sinpi(x) = sin(\pi \times x)$. Then, it assigns the common genetic algorithm setup, with a population size of **Pop=20**, maximum number of generations of **Gen=100**, elitism of **Eli=1**, crossover probability of 100% (**PCx=1.0**) and two mutation rates (**PMut1=0.01**, for single bit mutation implementations, and **PMut2=0.10**, for individual mutations). The code assumes a one-point crossover and bit mutation (shown in Fig. 5.2) to generate the new offspring. The one-point crossover is set by using the **recCrossover** (**ecr**) and **gabin_spCrossover** (**GA**) functions. Next, the different genetic algorithm implementations are executed, using the functions: **rbga.bin()**, **ecr()**, **ga()** and **GAopt()**. Each algorithm function contains its own arguments. Some of these arguments denote the same elements but use a different terminology. For instance, **rbga.bin()** uses **popSize** to define the population size, while **ecr** uses the **mu** argument for the same purpose. In case of **ecr**, the **lambda** argument defines how many new solutions are created in each generation. The selection operators for the **ecr** and **GA** packages are explicitly set by calling the **selRoulette** (**ecr**) and **gabin_nlrSelection** (**GA**) functions. After executing the algorithms, the code shows the best solution and evaluation value. An example execution output of file **maxsin-ga.R** is:

```
> source("maxsin-ga.R")
rbga.bin:
best: 0 1 1 1 1 1 1 1 1 1 1 1 1 1 1 1 1 1 1 1 f: 1
ecr:
best: 0 1 1 1 1 1 1 1 0 1 1 1 1 0 1 1 0 1 1 0 f: 0.9999798
ga:
best: 1 0 0 0 0 0 0 0 0 0 0 0 0 0 0 0 0 0 0 0 f: 1
GAopt:
best: 1 0 0 0 0 0 0 0 0 0 0 0 0 0 0 0 0 0 0 0 f: 1
```

All algorithms have obtained a value equal or very close to the perfect 1.0 evaluation. It should be noted that while the perfect solution is (1,0,...,0), some algorithm executions evolved a best solution (0,1,1,...,1) that obtains the same 1.0 **max sin** value. This occurs because in R all these instructions return the same value of **`1.0`**: **`sinpi(0.5)`**, **`sinpi(0.4999)`**, **`sin(pi*0.5)`** and **`sin(pi*0.4999)`**.

Regarding a continuous optimization example, the **rastrigin** task (with $D = 30$) is adopted in code **`rastrigin-ga.R`**:

```
### rastrigin-ga.R file ###

library(genalg)          # load rba.bin
library(ecr)        # load ecr
library(GA)         # load ga
library(mcga)       # load mcga

# real value rastrigin function with a dimension of 30:
rastrigin=function(x) 10*length(x)+sum(x^2-10*cos(2*pi*x))
rastrigin2=function(x) -rastrigin(x) # maximization goal
D=30

# common setup for the executions:
lower=rep(-5.2,D);upper=rep(5.2,D) # lower and upper bounds
Pop=20 # population size
Gen=100 # maximum number of generations
Eli=1 # elitism of one individual

set.seed(12345) # set for replicability
Dig=2 # use 2 digits to show results

# simple function to show best solution and evaluation value:
showres=function(s,f,digits=Dig)
 { cat("best:",round(s,Dig),"f:",round(f,Dig),"\n") }

# rbga (minimization):
cat("rbga:\n")
ga1=rbga(lower,upper,popSize=Pop,iters=Gen,
         evalFunc=rastrigin,elitism=Eli)
b=which.min(ga1$evaluations) # best individual
# show best solution and evaluation value:
showres(ga1$population[b,],ga1$evaluations[b])

# ecr (minimization):
cat("ecr:\n")
ga2=ecr(fitness.fun=rastrigin,minimize=TRUE,n.objectives=1,
        n.dim=D,
        lower=lower,upper=upper, # lower and upper bounds
        representation="float", # real values
        mu=Pop, # population size
        lambda=round(Pop/2), # half the population size
        n.elite=Eli, # elitism
        # tournament with k=2
        parent.selector=setup(selTournament,k=2),
        survival.selector=setp(selTournament,k=2),
```

```
        recombinator=recUnifCrossover, # uniform crossover
        mutator=setup(mutGauss,lower=lower,upper=upper),
        terminators=list(stopOnIters(Gen)))
showres(ga2$best.x[[1]],ga2$best.y)

# ga (maximization):
cat("ga:\n")
ga3=ga(type="real-valued",fitness=rastrigin2,
       lower=lower,upper=upper, # lower and upper bounds
       selection=ga_tourSelection, # tournament with k=3
       crossover=gabin_uCrossover, # uniform crossover
       mutation=gareal_raMutation, # uniform random mutation
       popSize=Pop, # population size
       elitism=Eli,maxiter=Gen,monitor=FALSE)
showres(ga3@solution,-ga3@fitnessValue)

# mcga (minimization):
cat("mcga:\n") # uniform binary crossover and bit mutation
ga4=mcga(popsize=Pop,chsize=D,elitism=Eli,
         minval=lower,maxval=upper,
         maxiter=Gen,evalFunc=rastrigin)
showres(ga4$population[1,],ga4$costs[1])
```

Similarly to the **max sin** demonstration, the code defines two rastrigin functions, where **rastrigin2** is used for the **ga** maximization goal. Then, the code sets a common setup: **lower** and **upper** bounds, population size of **Pop=20**, maximum generations of **Gen=100** and elitism of one **Eli=1**. Next, a simple auxiliary **showres()** function is set to present the results with a precision of **Dig=2** digits. The code includes four evolutionary algorithms: **rbga()**, **ecr()**, **ga()** and **mcga()**. As previously mentioned, the **rbga** and **mcga** packages are rather rigid and assume a fixed evolution algorithm, with a specific set of breeding and selection operators. For tutorial purposes, thus not attempting to achieve a similar execution, different operators were set for **ecr** and **ga**. The **rbga()** algorithm uses the default one-point crossover and the previously described real value mutation (with a random direction and dampening factor). The **ecr()** algorithm is set with a tournament selection ($k = 2$, configured via the **setup** and **selTournament** function), a uniform crossover (via the **recUnifCrossover** function), and a Gaussian mutation (via the **setup** and **mutGauss** functions). The Gaussian mutation adds a $\mathcal{N}(0, 1)$ generated noise to each real value. In this demonstration, the **ecr()** function was set with **lambda=round(Pop/2)**, which means that each generation creates an offspring that corresponds to half the population size. As for **ga()**, from the **GA** package, the execution uses a tournament selection (with default $k = 3$, via the **ga_tourSelection** function), a uniform crossover (via **gabin_uCrossover** function), and a uniform random mutation (via the **gareal_raMutation** function, which randomly sets a value within $[lower, upper]$). The function returns an S4 class instance, which is composed of slots that can be assessed via the **@** symbol (see Sect. 5.6). Finally, the **mcga()** uses an internal bit representation of the real values, thus new solutions are created via binary genetic operators, namely a uniform binary crossover and bit mutation. The package uses a fixed tournament selection with $k = 2$ as the selection operator. It should be noted that the **mcga** package only uses the lower and upper bounds to set the initial population but not the optimized parameters. This limitation does not affect the **rastrigin** demonstration, since the best solution is set at middle of the search range. However, if the optimized **mcga** values need to be bounded, then a constraint handling strategy (Sect. 1.5) should be adopted (e.g., death-penalty).

The output result of executing the **rastrigin** code is:

```
> source("rastrigin-ga.R")
rbga:
best: -0.87 -0.01 1.02 0.07 -0.11 -0.08 0.16 -1.05 0.99 -0.94
    -2.92 1.02 -0.02 1.99 -0.15 -1.14 0.05 -0.08 -2 0.06 -0.85 1
    .24 2.02 1.02 -2 1.05 1.97 1.89 -0.96 -1.09 f: 89.56
ecr:
best: -3.05 2 -1.99 0.99 -3 -0.93 0.97 -0.98 -0.91 0.05 1.02
    -1.98 -1.02 -0.99 -2.09 -3.07 -1.03 -1.92 -1 -1.05 -3.11 0 0
    .99 -1.01 0.87 3.06 0.96 -0.97 1.02 1.09 f: 101.1
ga:
best: -0.12 2.79 -0.99 -0.98 2.14 -3.13 -0.99 -0.88 -1.03 -0.9 1
    .15 -0.09 -0.92 -2.95 -1.08 -2 -1.74 2.23 1.1 -0.06 -1.99
    -2.99 1.84 -0.02 1.07 -0.06 1.12 -0.06 1.22 2.13 f: 151.51
mcga:
best: 0 0 0 0 0 0 0 0 0 0 0 0 0 0 0 0 0 0 0 0 0 0 0 0 0 0 0 0 0 0
     f: 0
```

In this case, the best solution was optimized using the **mcga** algorithm, which obtained an impressive perfect solution (0,0,...) and evaluation value (0.0). Nevertheless, it should be noted that the presented code is for tutorial purposes, to explain how the distinct packages can be used, and not to compare the algorithms (Sect. 5.7 exemplifies how to properly compare population based methods).

5.3 Differential Evolution

Differential evolution is a global search strategy that tends to work well for continuous numerical optimization and that was proposed in Storn and Price (1997). Similarly to genetic and evolutionary algorithms, the method evolves a population of solutions, where each solution is made of a string of real values. The main difference when compared with evolutionary algorithms is that differential evolution uses arithmetic operators to generate new solutions, instead of the classical crossover and mutation operators. The differential mutation operators are based on vector addition and subtraction, thus only working in metric spaces (e.g., Boolean, integer, or real values) (Luke, 2015).

This chapter details first the differential evolution algorithm implemented by the **DEoptim** package (Mullen et al., 2011), which performs a minimization goal and that also allows a parallel execution. The respective pseudo-code is presented in Algorithm 6.

Algorithm 6 Differential evolution algorithm as implemented by the **DEoptim** package

1: **Inputs:** f, C ▷ f is the evaluation (fitness) function, C includes control parameters
2: $P \leftarrow initialization(C)$ ▷ set initial population
3: $B \leftarrow best(P, f)$ ▷ best solution of the initial population
4: $i \leftarrow 0$ ▷ i is the number of iterations of the method
5: **while** not $termination\_criteria(P, f, C, i)$ **do** ▷ DEoptim uses up to three termination criteria
6: **for** each individual $s \in P$ **do** ▷ cycle all population individuals
7: $s' \leftarrow mutation(P, C)$ ▷ differential mutation, uses parameters F and CR
8: **if** $f(s') < f(s)$ **then** $P \leftarrow replace(P, s, s')$ ▷ replace s by s' in the population
9: **end if**
10: **if** $f(s') < f(B)$ **then** $B \leftarrow s'$ ▷ minimization goal
11: **end if**
12: **end for**
13: $i \leftarrow i + 1$
14: **end while**
15: **Output:** B, P ▷ best solution and last population

The classical differential mutation starts by first choosing three individuals (s_1, s_2, and s_3) from the population. In contrast with genetic algorithms, these three individuals are randomly selected and selection only occurs when replacing mutated individuals in the population (as shown in Algorithm 6). A trial mutant is created using (Mullen et al., 2011): $s_{m,j} = s_{1,j} + F \times (x_{2,j} - x_{3,j})$, where $F \in [0, 2]$ is a positive scaling factor, often less than 1.0, and j denotes the j-th parameter of the representation of the solution. If the trial mutant values violate the upper or lower bounds, then $s_{m,j}$ is reset using $s_{m,j} = \max(s_j) - \mathcal{U}(0, 1)(\max(s_j) - min(s_j))$, if $s_{m,j} > \max(s_j)$, or $s_{m,j} = \min(s_j) + \mathcal{U}(0, 1)(\max(s_j) - \min(s_j))$, if $s_{m,j} < \min(s_j)$, where $\max(s_j)$ and $\min(s_j)$ denote the upper and lower limits for the j-th parameter of the solution. The first trial mutant value (chosen at random) is always computed. Then, new mutations are generated until all string values have been mutated (total of L_S mutations) or if $r > CR$, where $r = \mathcal{U}(0, 1)$ denotes a random number and CR is the crossover probability. Finally, the new child (s') is set as the generated mutant values plus the remaining ones from the current solution (s). Hence, the CR constant controls the fraction of values that are mutated.

The **DEoptim** function is coherent with the known **optim** R function, and it includes the arguments:

- **fn**—function to be minimized;
- **lower**, **upper**—lower and upper bounds;
- **control**—a list of control parameters (details are given in function **DEoptim.control**);
- **...**—additional arguments to be passed to **fn**; and
- **fnMap**—optional function that is run after each population creation but before the population is evaluated (it allows to impose integer/cardinality constraints).

The control parameters (C) are specified using the **DEoptim.control** function, which contains the arguments such as:

- **VTR**—value to be reached, stop if best value is below VTR (default **VTR=-Inf**);
- **strategy**—type of differential strategy adopted, includes six different mutation strategies (classical mutation is set using **strategy=1**, default is **strategy=2**, full details can be accessed by executing `> ?DEoptim.control`);
- **NP**—population size (default is **10*length(lower)**);
- **itermax**—maximum number of iterations (default is **200**);
- **CR**—crossover probability ($CR \in [0, 1]$, default is **0.5**);
- **F**—differential weighting factor ($F \in [0, 2]$, default is **0.8**);
- **trace**—logical or integer value indicating if progress should be reported (if integer it occurs every **trace** iterations, default is **true**);
- **initialpop**—an initial population (defaults to **NULL**);
- **storepopfrom**, **storepopfreq**—from which iteration and with which frequency should the population values be stored; and
- **reltol**, **steptol**—relative convergence tolerance stop criterion, the method stops if unable to reduce the value by a factor of **reltol*(abs(value))** after **steptol** iterations.

The result of the **DEoptim** function is a list that contains 2 components:

- **$optim**—a list with elements, such as **$bestmem**—best solution and **$bestval**—best evaluation value;
- **$member**—a list with components, such as **bestvalit**—best value at each iteration and **pop**—last population.

Similarly to the **genalg** package, **DEoptim** also includes functions for plotting (**plot.DEoptim**) and summarizing (**summary.DEoptim**) results (under S3 scheme of method dispatching).

Price et al. (2005) advise the following general configuration for the differential evolution parameters: use the default $F = 0.8$ and $CR = 0.9$ values and set the population size to ten times the number of solution values ($N_P = 10 \times L_S$). Further details about the **DEoptim** package can be found in Mullen et al. (2011) (execute `> vignette("DEoptim")` to get an immediate access to this reference).

The demonstration **sphere-DEoptim.R** code adopts the **sphere** ($D = 2$) task:

```
### sphere-DEoptim.R file ###
library(DEoptim) # load DEoptim

sphere=function(x) sum(x^2)
D=2
maxit=100
set.seed(12345) # set for replicability
C=DEoptim.control(strategy=1,NP=5,itermax=maxit,CR=0.9,F=0.8,
                  trace=25,storepopfrom=1,storepopfreq=1)
# perform the optimization:
```

```
D=suppressWarnings(DEoptim(sphere,rep(-5.2,D),rep(5.2,D),
                           control=C))
# show result:
summary(D)
pdf("DEoptim.pdf",onefile=FALSE,width=5,height=9,
    colormodel="gray")
plot(D,plot.type="storepop")
dev.off()
cat("best:",D$optim$bestmem,"f:",D$optim$bestval,"\n")
```

The **C** object contains the control parameters, adjusted for the classical differential mutation and population size of 5, among other settings (the arguments **storepopfrom** and **storepopfreq** are required for the **plot**). Giving that **DEoptim** produces a warning when the population size is not set using the advised rule ($N_P = 10 \times L_S$), the **suppressWarnings** R function was added to ignore such warning. Regarding the plot, the informative **"storepop"** was selected (other options are **"bestmemit"**—evolution of the best parameter values; and **"bestvalit"**—best function value in each iteration). Also, additional arguments were used in the **pdf** function (**onefile** and **colormodel**) in order to adjust the file to contain just one page with a gray coloring scheme. The execution result of file **sphere-DEoptim.R** is:

```
> source("sphere-DEoptim.R")
Iteration: 25 bestvalit: 0.644692 bestmemit:    0.799515    0
    .073944
Iteration: 50 bestvalit: 0.308293 bestmemit:    0.550749
    -0.070493
Iteration: 75 bestvalit: 0.290737 bestmemit:    0.535771
    -0.060715
Iteration: 100 bestvalit: 0.256731 bestmemit:    0.504867
    -0.042906

***** summary of DEoptim object *****
best member    :  0.50487 -0.04291
best value     :  0.25673
after          :  100 generations
fn evaluated   :  202 times
*************************************
best: 0.5048666 -0.0429055 f: 0.2567311
```

The differential evolution algorithm improved the best value of the initial population from 7.37 to 0.25, leading to the optimized solution of $s = (0.50, -0.04)$. Figure 5.6 presents the result of the plot, showing a fast convergence in the population toward the optimized values (0.5 and −0.04). The first optimized value (0.5) is not very close to the optimum. However, this is a tutorial example that includes a very small population size. When the advised rule is used ($N_P = 20$), the maximum distance of the optimized point to the origin is very small (7.53×10^{-10}).

The CRAN task view for Optimization (https://cran.r-project.org/web/views/Optimization.html) lists two additional packages that implement a differential evolution: **DEoptimR** (Conceicao, 2016) and **NMOF** (Gilli et al., 2019). The pre-

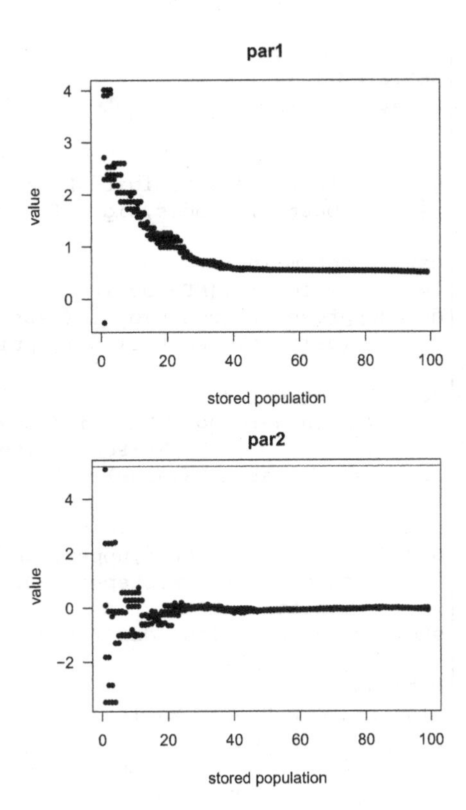

Fig. 5.6 Population evolution in terms of x_1 (top) and x_2 (bottom) values under the differential evolution algorithm for **sphere** ($D = 2$)

viously described **GA** package also includes a differential evolution function (**de**). For tutorial purposes, the demonstration file **rastrigin-de.R** runs the four differential evolution algorithms, implemented in the **DEoptim()**, **JDEoptim()**, **de()**, and **DEopt()** functions, for the **rastrigin** task ($D = 30$):

```
### rastrigin-de.R file ###

library(DEoptim)  # load DEoptim
library(DEoptimR) # load JDEoptim
library(GA)       # load GA
library(NMOF)     # load DEopt

# real value rastrigin function with a dimension of 30:
rastrigin=function(x) 10*length(x)+sum(x^2-10*cos(2*pi*x))
rastrigin2=function(x) -rastrigin(x) # maximization goal
D=30

# common setup for the executions:
lower=rep(-5.2,D);upper=rep(5.2,D) # lower and upper bounds
Pop=20 # population size
Maxit=100 # maximum number of iterations
PCx=0.9 # fraction of values that are mutated
Df=0.8 # differential weighting factor
```

```
set.seed(12345) # set for replicability
Dig=2 # use 2 digits to show results

# simple function to show best solution and evaluation value:
showres=function(s,f,digits=Dig)
 { cat("best:",round(s,Dig),"f:",round(f,Dig),"\n") }

cat("DEoptim:\n") # minimization
C=DEoptim.control(NP=Pop,itermax=Maxit,CR=PCx,F=Df,trace=FALSE)
de1=suppressWarnings(DEoptim(rastrigin,lower,upper,control=C))
showres(de1$optim$bestmem,de1$optim$bestval)

cat("JDEoptim:\n") # minimization
de2=suppressWarnings(JDEoptim(lower,upper,fn=rastrigin,
                     NP=Pop,maxiter=Maxit))
showres(de2$par,de2$value)

cat("de:\n") # maximization
de2=de(rastrigin2,lower,upper,popSize=Pop,maxiter=Maxit,
       pcrossover=PCx,stepsize=Df,
       monitor=FALSE)
showres(de2@solution,-de2@fitnessValue)

cat("DEopt:\n") # minimization
algo=list(nP=Pop,nG=Maxit,
          CR=PCx, # crossover probability
          F=Df, # default step size
          min=lower,max=upper, # lower and upper bounds
          printDetail=FALSE,printBar=FALSE)
de4=DEopt(rastrigin,algo=algo)
showres(de4$xbest,de4$OFvalue)
```

Three of the implementations (**DEoptim**, **de**, and **NMOF**) allow the definition of similar parameters, namely the fraction of mutated values (**PCx=0.9** in this demonstration) and differential weighting factor (**Df=0.8**). The other experimentation parameters were set as **Pop=20** (population size) and **Maxit=100** (stopping criterion). Both **DEoptim()** and **JREoptim()** return execution warnings and thus the **suppressWarnings()** R function was used to remove them from the console. The **rastrigin-de.R** file execution result is:

```
> source("rastrigin-de.R")
DEoptim:
best: -0.97 -0.58 -1.14 0.02 0.23 -0.56 1.09 0.18 0.03 -0.05
    -1.25 -0.12 1 -0.07 -0.97 2.28 -0.03 -0.84 -1.08 -3.63 -2.17
    0.08 -0.15 1.87 -0.88 2 -1.1 0 0.89 -0.95 f: 172.03
JDEoptim:
best: -1.06 0.96 -1.09 -0.07 0.78 0.75 2.01 0.07 -0.14 -0.13
    -1.66 -1.16 1.06 -0.35 -0.08 0.12 -0.16 0.04 1.11 -1.05 -4.05
     1.11 -1.09 1.04 -0.75 -2.98 -1.9 1.16 -0.76 0.97 f: 160.41
de:
```

```
best: 3.06 -1.1 -1.24 -2.27 3.06 -1.33 0 1.46 -1.9 -3.16 -0.1
    -1.99 -0.73 0.1 -0.89 1.96 1.91 1.54 -2.97 1.98 -0.56 0.79 3
    .47 -1.34 0.42 -3.04 -0.06 -0.98 -3.83 -1.57 f: 324.35
DEopt:
best: 0.31 1.49 1.22 0.13 1.96 1.11 -0.84 -1.45 1.38 2.21 3.11
    -2.69 -0.52 3.82 1.7 -2.64 0.14 -5.45 1.1 -0.03 -1.29 0.24 1
    .27 -2.5 0.79 -1.86 -1.89 0.9 5.19 -0.12 f: 406.37
```

In this execution, the best result was achieved by **JDEoptim** (evaluation value of 160.41), followed by **DEOptim** (evaluation of 172.03).

5.4 Particle Swarm Optimization

Swarm intelligence denotes a family of algorithms (e.g., ant colony and particle swarm optimization) that are inspired in swarm behavior, which is exhibited by several animals (e.g., birds, ants, bees). These algorithms assume a population of simple agents with direct or indirect interactions that influence future behaviors. While each agent is independent, the whole swarm tends to produce a self-organized behavior, which is the essence of swarm intelligence (Michalewicz et al., 2006).

Particle swarm optimization is a swarm intelligence technique that was proposed by Kennedy and Eberhart (1995) for numerical optimization. Similarly to differential evolution, particle swarms operate mostly on metric spaces (Luke, 2015). The algorithm is defined by the evolution of a population of particles, represented as vectors with a D-th (or L_S) dimensional space. The particle trajectories oscillate around a region that is influenced by the previous performance of the particle and by the success of its neighborhood (Mendes et al., 2002).

Since the original algorithm was presented in 1995, numerous variants have been proposed. This chapter details first the **pso** package, which implements two standard versions that have been made publicly available at the Particle Swarm Central site (http://www.particleswarm.info/): SPSO 2007 and 2011. It is importance to note that these SPSO variants do not claim to be the best versions on the market. Rather, SPSO 2007 and 2011 implement the original particle swarm version (Kennedy and Eberhart, 1995) with a few improvements. The goal is to define stable standards that can be compared against newly proposed particle swarm algorithms.

A particle moves on a step by step basis, where each step is also known as iteration, and contains (Clerc, 2012): a position (s, inside search space), a fitness value (f), a velocity (v, used to compute next position), and a memory (p, previous best position found by the particle, and l, previous best position in the neighborhood). Each particle starts with random position and velocity values. Then, the search is performed by a cycle of iterations. During the search, the swarm assumes a topology, which denotes a set of links between particles. A link allows one particle to inform another one about its memory. The neighborhood is defined by the set of informants of a particle. The new particle position depends on its

Algorithm 7 Particle swarm optimization pseudo-code for SPSO 2007 and 2011

```
1: Inputs: f, C                       ▷ f is the evaluation function, C includes control parameters
2: P ← initialization(C)              ▷ set initial swarm (topology, random position and velocity,
   previous best and previous best position found in the neighborhood)
3: B ← best(P, f)                                                         ▷ best particle
4: i ← 0                                        ▷ i is the number of iterations of the method
5: while not termination_criteria(P, f, C, i) do
6:     for each particle x = (s, v, p, l) ∈ P do                      ▷ cycle all particles
7:         v ← velocity(s, v, p, l)                               ▷ compute new velocity for x
8:         s ← s + v                          ▷ move the particle to new position s (mutation)
9:         s ← confinement(s, C)                  ▷ adjust position s if it is outside bounds
10:        if f(s) < f(p) then p ← s                                ▷ update previous best
11:        end if
12:        x ← (s, v, p, l)                                              ▷ update particle
13:        if f(s) < f(B) then B ← s                                   ▷ update best value
14:        end if
15:    end for
16:    i ← i + 1
17: end while
18: Output: B                                                             ▷ best solution
```

current position and velocity, while velocity values are changed using all elements of a particle (s, v, p, and l). The overall pseudo-code for SPSO 2007 and 2011 is presented in Algorithm 7, which assumes a minimization goal.

This chapter highlights only the main SPSO differences. The full details are available at Clerc (2012). Several termination criteria are defined in SPSO 2007 and 2011: maximum admissible error (when optimum point is known), maximum number of iterations/evaluations, and maximum number of iterations without improvement. Regarding the swarm size (N_P), in SPSO 2007 it is automatically defined as the integer part of $10 + 2\sqrt{L_S}$ (L_S denotes the length of a solution), while in SPSO 2011 it is user defined (with suggested value of 40).

Historically, two popular particle swarm topologies (Mendes et al., 2002) were: *star*, where all particles know each other; and *ring*, where each particle has only 2 neighbors. However, the SPSO variants use a more recent adaptive star topology (Clerc, 2012), where each particle informs itself and K randomly particles. Usually, K is set to 3. SPSO 2007 and 2011 use similar random uniform initializations for the position: $s_j = \mathcal{U}(\min(s_j), \max(s_j)), j \in \{1, \ldots, L_S\}$. However, the velocity is set differently: SPSO 2007—$v_j = \frac{\mathcal{U}(\min(s_j), \max(s_j)) - s_j}{2}$; SPSO 2011—$v_j = \mathcal{U}(\min(s_j) - s_j, \max(s_j) - s_j)$.

In SPSO 2007, the velocity update is applied dimension by dimension:

$$v_j \leftarrow wvj + \mathcal{U}(0, c)(p_j - s_j) + \mathcal{U}(0, c)(l_j - s_j) \tag{5.1}$$

where w and c are exploitation and exploration constants. The former constant sets the ability to explore regions of the search space, while the latter one defines the ability to concentrate around a promising area. The suggested values for these constants are $w = 1/(2\ln 2) \simeq 0.721$ and $c = 1/2 + \ln(2) \simeq 1.193$. A different scheme is adopted for SPSO 2011. First, the center of gravity (G_j) of three points (current position and two points, slightly beyond p and l):

$$G_j = s_j + c\frac{p_j + l_j - 2s_j}{3} \tag{5.2}$$

Then, a random point (s') is selected within the hypersphere of center G_j and radius $\| G_j - s_j \|$. Next, the velocity is updated as $v_j = wvj + s'_j - s_j$ and the new position is simply adjusted using $s_j = wv_j + s'_j$. There is a special case, when $l = p$. In such case, in SPSO 2007, the velocity is set using $v_j \leftarrow wvj + \mathcal{U}(0, c)(p_j - s_j)$, while in SPSO 2011 the center of gravity is computed using $s_j + c\frac{p_j - s_j}{2}$.

The *confinement* function is used to assure that the new position is within the admissible bounds. In both SPSO variants, when a position is outside a bound, it is set to that bound. In SPSO 2007, the velocity is also set to zero, while in SPSO 2011 it is set to half the opposite velocity ($v_j = -0.5v_j$).

The **pso** package includes the core **psoptim** function that can be used as a replacement of function **optim** (Sect. 4.3) and includes arguments, such as:

- **par**—vector defining the dimensionality of the problem (L_S), included for compatibility with optim and can include **NA** values;
- **fn**—function to be minimized;
- **lower**, **upper**—lower and upper bounds;
- **...**—additional arguments to be passed to **fn**; and
- **control**—a list of control parameters.

The control list includes components, such as:

- **$trace**—if positive, progress information is shown (default is 0);
- **$fnscale**—scaling applied to the evaluation function (if negative, transforms the problem into maximization; default is 1);
- **$maxit**—maximum number of iterations (defaults to 1000);
- **$maxf**—maximum number of function evaluations;
- **$abstol**—stops if best fitness is less than or equal to this value (defaults to **-Inf**);
- **$reltol**—if the maximum distance between best particle and all others is less than **reltol*d**, then the algorithm restarts;
- **$REPORT**—frequency of reports if **trace** is positive;
- **$trace.stats**—if **TRUE**, then statistics at every **REPORT** step are recorded;
- **$s**—swarm size ($N_P$);
- **$k**—$K$ value (defaults to 3);
- **$p**—average percentage of informants for each particle (a value of 1 implies a fully informed scheme, where all particles and not just K neighbors affect the individual velocity, defaults to **1-(1-1/s)^k**);
- **$w**—exploitation constant (if a vector of 2 elements, constant is gradually changed from **w[1]** to **w[2]**, default **1/(2*log(2))**);

- **$c.p**—local exploration constant (associated with p, defaults to **0.5+log(2)**);
- **$c.g**—global exploration constant (associated with l, defaults to **0.5+log(2)**);
- **$d**—diameter of the search space (defaults to Euclidean distance between **upper** and **lower**);
- **$v.max**—maximum admitted velocity (if not **NA** the velocity is clamped to the length of **v.max*d**, defaults to **NA**);
- **$maxit.stagnate**—maximum number of iterations without improvement (defaults to **Inf**); and
- **$type**—SPSO implementation type ("SPSO2007" or "SPSO2011," defaults to "SPSO2007").

The result is a list (compatible with **optim**) that contains:

- **$par**—best solution found;
- **$value**—best evaluation value;
- **$counts**—vector with three numbers (function evaluations, iterations, and restarts);
- **$convergence** and **$message**—stopping criterion type and message; and
- **$stats**—if **trace** is positive and **trace.stats** is true then contains the statistics: **it**—iteration numbers, **error**—best fitness, **f**—current swarm fitness vector, and **x**—current swarm position matrix.

The **sphere-psoptim.R** file adapts the **psoptim** function for the **sphere** ($D = 2$) task:

```
### sphere-psoptim.R file ###
library(pso) # load pso

sphere=function(x) sum(x^2)

D=2; maxit=10; s=5
set.seed(12345) # set for replicability
C=list(trace=1,maxit=maxit,REPORT=1,trace.stats=1,s=s)
# perform the optimization:
PSO=psoptim(rep(NA,D),fn=sphere,lower=rep(-5.2,D),
            upper=rep(5.2,D),control=C)
# result:
pdf("psoptim1.pdf",width=5,height=5)
j=1 # j-th parameter
plot(xlim=c(1,maxit),rep(1,s),PSO$stats$x[[1]][j,],pch=19,
     xlab="iterations",ylab=paste("s_",j," value",sep=""))
for(i in 2:maxit) points(rep(i,s),PSO$stats$x[[i]][j,],pch=19)
dev.off()
pdf("psoptim2.pdf",width=5,height=5)
plot(PSO$stats$error,type="l",lwd=2,xlab="iterations",
     ylab="best fitness")
dev.off()
cat("best:",PSO$par,"f:",PSO$value,"\n")
```

In this demonstration, a very small swarm size ($N_P = 5$) was adopted. Also, the control list was set to report statistics every iteration, under a maximum of

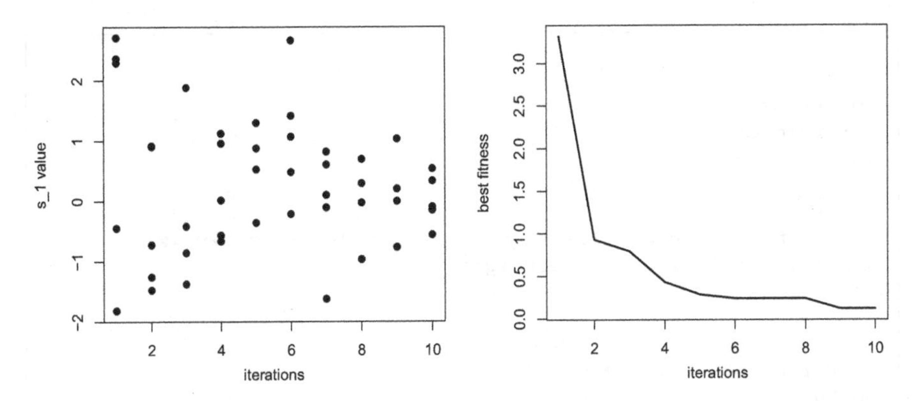

Fig. 5.7 Particle swarm optimization for **sphere** and $D = 2$ (left denotes the evolution of the position particles for the first parameter; right shows the evolution of the best fitness)

10 iterations. The visual results are presented in terms of two plots. The first plot is similar to Fig. 5.6 and shows the evolution of the position particles for the first parameter. The second plot shows the evolution of the best fitness during the optimization. The execution of file **sphere-psoptim.R** is:

```
> source("sphere-psoptim.R")
S=5, K=3, p=0.488, w0=0.7213, w1=0.7213, c.p=1.193, c.g=1.193
v.max=NA, d=14.71, vectorize=FALSE, hybrid=off
It 1: fitness=3.318
It 2: fitness=0.9281
It 3: fitness=0.7925
It 4: fitness=0.4302
It 5: fitness=0.2844
It 6: fitness=0.2394
It 7: fitness=0.2383
It 8: fitness=0.2383
It 9: fitness=0.1174
It 10: fitness=0.1174
Maximal number of iterations reached
best: 0.2037517 -0.2755488 f: 0.1174419
```

The particle swarm improved the best fitness from 3.318 to 0.1174, leading to the optimized solution of $s = (0.20, -0.28)$ ($f = 0.12$). Figure 5.7 presents the result of the plots. It should be stressed that this tutorial example uses a very small swarm. When the advised rule is adopted ($N_P = 12$), the optimized values are $f = 0.03$ (10 iterations) and $f = 9.36 \times 10^{-12}$ (100 iterations).

There are other particle swarm R implementations, such as provided by the **psoptim** (Ciupke, 2016) and **NMOF** (Gilli et al., 2019) packages. Similarly to the previous sections, these distinct implementations are demonstrated with the **rastrigin** task ($D = 30$) in file **rastrigin-pso.R**:

```
### rastrigin-pso.R file ###

library(pso)        # load pso::psoptim
library(psoptim)    # load psoptim::psoptim
library(NMOF)       # load PSopt

# real value rastrigin function with a dimension of 30:
rastrigin=function(x) 10*length(x)+sum(x^2-10*cos(2*pi*x))
rastrigin2=function(x) -rastrigin(x)
D=30

# common setup for the executions:
lower=rep(-5.2,D);upper=rep(5.2,D) # lower and upper bounds
Pop=20 # swarm size
Maxit=100 # maximum number of iterations

set.seed(12345) # set for replicability
Dig=2 # use 2 digits to show results

# simple function to show best solution and evaluation value:
showres=function(s,f,digits=Dig)
 { cat("best:",round(s,Dig),"f:",round(f,Dig),"\n") }

cat("pso::psoptim:\n") # minimization
C=list(maxit=Maxit,s=Pop,type="SPSO2011") # SPSO2011
pso1=pso::psoptim(rep(NA,D),fn=rastrigin,
                  lower=lower,upper=upper,control=C)
showres(pso1$par,pso1$value)

cat("psoptim::psoptim:\n") # maximization
pso2=psoptim::psoptim(FUN=rastrigin2,
                      n=Pop, # swarm size
                      max.loop=Maxit,
                      xmin=lower,xmax=upper,
                      #velocity constraints in each direction:
                      vmax=rep(4,D), # default 4 value is used
                      anim=FALSE)
showres(pso2$sol,-pso2$val)

cat("PSopt:\n") # minimization
algo=list(nP=Pop,nG=Maxit,
          min=lower,max=upper, # lower and upper bounds
          printDetail=FALSE,printBar=FALSE)
pso3=PSopt(rastrigin,algo=algo)
showres(pso3$xbest,pso3$OFvalue)
```

The common setup assumes a swarm size of **Pop=20** and a stopping criterion of **Maxit=100**. Since there are two packages (**pso** and **psoptim**) that adopt the same function name, the two functions are distinguished by using the **::** R operator. For **pso::psoptim** (function of the **pso** package), the standard SPSO 2011 variant is selected. As for **psoptim::psoptim()**, the function arguments are not coherent with **optim()**. This function performs a maximization task, thus

FUN=rastrigin2. It also requires the setting of the **vmax** argument, a vector of size D with velocity constraints (set in the demonstration to **rep(4,D)**). Finally, the **PSopt()** function assumes the typical **NMOF** package usage.

The execution result of file **rastrigin-pso.R** is:

```
> source("rastrigin-pso.R")
pso::psoptim:
best: 3.97 -0.14 0.43 2.15 2.13 0.24 -0.03 -0.02 -0.06 -0.81 2.05
      0.97 4.07 2.11 1.21 1.04 0.11 0.92 2.19 -1.78 0.05 3.77 0.69
      -1.09 -0.82 1.05 -1.1 0.29 1.77 1.42 f: 237.42
psoptim::psoptim:
best: 2.13 4.23 -0.01 -3.25 3.67 -3.63 -0.09 1.53 1.23 -2.32
     -3.18 -4.31 -0.86 4.06 -1.01 1.56 0.05 -3.34 1.27 2.52 4.76
     -1.12 -5.06 2.81 0.88 3.09 -1.03 0.45 -2.07 -1.11 f: 446.65
PSopt:
best: 0.1 -2.29 0.99 -1.01 -0.92 0.17 -0.84 -1.95 0.04 0.25 -0.07
      -0.13 -1.04 0.13 0.88 0.08 0.7 -0.24 0 -0.99 -0.06 -0.03
     -1.21 0.09 0.06 -0.03 -0.03 1.21 -0.85 -0.78 f: 121.6
```

In this tutorial demonstration, the best **rastrigin** value (121.6) was obtained using the **NMOF** implementation.

5.5 Ant Colony Optimization

Ant colony optimization is a swarm intelligence that was proposed by Dorigo et al. (1996). The method is inspired by the foraging behavior of ants. Ants tend to randomly explore the area surrounding their nest. When food is found, the ant carries it back to the nest, leaving a pheromone trail. This pheromone guides other ants to the food source and it is reinforced if further food is found. There is also an evaporation mechanism that diminishes the pheromone trails over time. Thus, shorter food paths, used more frequently by ants, tend to have stronger pheromone trails.

Ant colony algorithms were initially applied to combinatorial optimization tasks, such as the Traveling Salesman Problem (TSP) (Sect. 7.2), scheduling and vehicle routing (Dorigo and Stützle, 2004). When using this optimization method, the goal needs to be formulated as finding the shortest path on a weighted graph. The algorithm search is guided by two mechanisms (Yang, 2014): a random route generation, which is a fitness-proportional mutation, and a pheromone-based (fitness) selection, which promotes shorter routes. The generic algorithm is presented in Algorithm 8. Within its main cycle, each ant constructs stochastically a solution (*ConstructSolution*) that is based on the order of the edges of the weighted graph. Each candidate solution is termed an ant trail. As for the *PheromoneUpdate()*, it can be based on local and global updates. The aim is to reinforce pheromone values ($\vec{p}$) associated with good solutions and decrease low quality ant trails (evaporation).

In (Socha and Dorigo, 2008), the method was also adapted for continuous domains by considering the search space as a finite set of continuous decision variables. The population P consists of N_P ant trails that are incrementally

Algorithm 8 Generic ant colony pseudo-code

```
1: Inputs: f, C                ▷ f is the evaluation function, C includes control parameters
2: i ← 0                       ▷ i is the number of iterations of the method
3: p⃗ ← InitializePheromone(C)
4: while not termination_criteria(f, C, i) do
5:     P ← ConstructSolution(p⃗, C)                  ▷ construct the population
6:     F_P ← f(P)                                   ▷ evaluate current population
7:     B ← best(P, F_P, C)                          ▷ select best solution
8:     p⃗ ← PheromoneUpdate(p⃗, P, F_P, C)
9:     i ← i + 1
10: end while
11: Output: B                                       ▷ best solution
```

constructed based on a pheromone probabilistic choice, using a probability density function, namely the Gaussian kernel:

$$G^i(x) = \sum_{i=1}^{k} w_l \mathcal{N}(\mu_l^i, \sigma_l^i) \tag{5.3}$$

where $i \in \{1, \ldots, D\}$, $l \in \{1, \ldots, k\}$, $w = (w_1, \ldots, w_l)$ denotes a weight vector and $\mathcal{N}(\mu, \sigma)$ is the standard Gaussian function. The ant colony continuous extension assumes that there is an archive of solutions (T) of size k, which defines the complexity of the Gaussian kernel: $T = \{s_1, s_2, \ldots, s_k\}$, where $s_l = (s_l^1, s_l^2, \ldots, s_l^D)$ denotes a solution.

Each dimension $i \in \{1, \ldots, D\}$ has a Gaussian kernel G^i. All ith values in the archive solutions are the elements of $\mu^i = (s_l^i, s_l^i, \ldots, s_i^k)$. The solutions in the archive T are ranked, thus s_l has rank l. The weight w_l is computed as:

$$w_l = \frac{1}{qk\sqrt{2\pi}} \exp\left(-\frac{(l-1)^2}{2q^2k^2}\right) \tag{5.4}$$

which is a Gaussian function with mean $\mu = 1.0$ and standard deviation $\sigma = qk$. The q parameter sets the locality of the search. If small q values are used, best ranked solutions are favored, else the search is more uniform.

In *ConstructSolution()*, an ant j constructs a solution $s_j = (s_j^1, \ldots, s_j^D)$ by executing D steps. First, the elements of the weight vector w are computed (Eq. (5.4)). Then, two samplings are executed: a Gaussian function is selected from the Gaussian kernel by using a probability $p_l = \frac{w_l}{\sum_{r=1}^{k} w_r}$; and a s_l^i value is generated according to the selected normal distribution. Given that μ_l^i is obtained using the archive T and w is already computed, the estimation of the s_l^i value requires the computation of σ_l^i. At each ith step, this value is set using the mean distance from solution s_l to the other solutions in the archive:

$$\sigma_l^i = \xi \sum_{e=1}^{k} \frac{|s_e^i - s_l^i|}{k-1} \tag{5.5}$$

where $\xi > 0$ is a parameter that produces an effect that is similar to the pheromone evaporation rate in the standard ant colony optimization. High ξ values reduce the convergence of the algorithm, making the search less biased toward previously explored solutions. After construction of the solutions, the pheromone information, which consists in the archive T, is updated (*PheromoneUpdate*). The update works by adding newly generated solutions to the archive and removing the worst solutions, such that the archive size (k) does not change.

The continuous ant colony algorithm is demonstrated with the **sphere** task ($D = 30$) by using the **evoper** package. The package assumes that the evaluation function receives D argument values, each with a real value (and not a single vector). The function needs to be set as a **PlainFunction** object (with pure R code), which also includes the D argument names, lower and upper bounds (these are called **parameters** in the package). As for the algorithm arguments, these are: *maxit*—**iterations**, N_P—**n.ants**, k—**k**, q—**q** and ξ—**Xi**. To set these arguments, the package requires the definition of an **OptionsACOR** object. Then, the algorithm can be executed by running the **abm.acor()** function. The demonstration code file **sphere-acor.R** is:

```
### sphere-acor.R file ###

library(evoper) # load evoper

# real value sphere function with a dimension of 30:
sphere=function(x) sum(x^2)
sphere2=function(...) # called as: sphere2(x1,x2,...,xD)
{ # ... means a variable sequence of arguments
 args=as.list(match.call()) # get the ... arguments
 # args: sphere2 and the ... arguments
 # args[[1]]="sphere2", thus use only 2:length(args)
 x=vector(length=length(args)-1)
 for(i in 2:length(args)) x[i-1]=as.numeric(args[[i]])
 return(sphere(x))
}
D=30 # dimension
lower=-5.2;upper=5.2 # lower and upper bounds
f=PlainFunction$new(sphere2)
# set the 30 parameters with lower and upper bounds:
for(i in 1:30)
  f$Parameter(name=paste("x",i,sep=""),min=lower,max=upper)

# set acor internal parameters:
opt=OptionsACOR$new() # get default acor options
opt$setValue("iterations",10)
opt$setValue("n.ants",64) # The number of simulated ants

set.seed(12345) # set for replicability

# run the ant colony optimization:
```

```
aco=abm.acor(f,opt)
b=aco$overall.best # best solution
print(b)
```

The auxiliary **sphere2** function converts the D arguments into a single vector **x** used by the standard **sphere()**. Since D is of variable size, the special three dots argument (**...**, Sect. 2.3) is used to capture all arguments. The line **args=as.list(match.call())** sets a vector list with all inputted arguments (including the function name, which is the first element of the list). Then, a **for** cycle is used to fill the **x** vector with the individual arguments, such that the value of **sphere(x)** can be returned. The **objective** argument of **abm.acor()** is set as object **f**, which includes the **sphere2** (as a **PlainFunction**) and the $D = 30$ parameters (with the names **"x1"**, **"x2"**, ..., **"x30"** and $[-5.2, 5.2]$ lower and upper bounds). Turning to the algorithm internal parameters, these are set by defining a **OptionsACOR** object. Firstly, the **opt=OptionsACOR$new()** instruction is used to get the default values. Then, the iterations and number of ants are set to $maxit = 10$ and $N_P = 64$, using the **opt$setValue** code. The algorithm is then called using the **abm.acor()** function and the best solution is stored in object **b**. The **sphere** demonstration result is:

```
> source("sphere-acor.R")
INFO [2020-06-26 14:02:00] Options(acor): iterations=10  trace=0
        n.ants=64  k=32   q=0.2    Xi=0.5
INFO [2020-06-26 14:02:00] Initializing solution
INFO [2020-06-26 14:02:00] Starting metaheuristic
INFO [2020-06-26 14:02:00] Iteration=1/10, best fitness=146.256,
    iteration best fitness=146.256
INFO [2020-06-26 14:02:01] Iteration=2/10, best fitness=146.256,
    iteration best fitness=151.26
INFO [2020-06-26 14:02:01] Iteration=3/10, best fitness=146.256,
    iteration best fitness=146.315
INFO [2020-06-26 14:02:01] Iteration=4/10, best fitness=122.231,
    iteration best fitness=122.231
INFO [2020-06-26 14:02:01] Iteration=5/10, best fitness=114.53,
    iteration best fitness=114.53
INFO [2020-06-26 14:02:01] Iteration=6/10, best fitness=106.101,
    iteration best fitness=106.101
INFO [2020-06-26 14:02:01] Iteration=7/10, best fitness=92.4889,
    iteration best fitness=92.4889
INFO [2020-06-26 14:02:01] Iteration=8/10, best fitness=85.703,
    iteration best fitness=85.703
INFO [2020-06-26 14:02:01] Iteration=9/10, best fitness=74.5561,
    iteration best fitness=74.5561
INFO [2020-06-26 14:02:01] Iteration=10/10, best fitness=68.3807,
     iteration best fitness=68.3807
           x1         x2          x3         x4         x5         x6
                          x7          x8         x9        x10        x11
                        x12        x13        x14
52 -0.5816636 3.272893 -0.302876 0.4033882 0.3475002 0.8233158
    -0.8289073 -0.7933312 0.8626004 -2.065832 1.364325 1.722329 3
    .643104 -2.988287
```

```
         x15       x16        x17        x18       x19        x20
                    x21        x22        x23         x24        x25
                   x26          x27         x28
52 0.5921534 1.526668 -1.200542 0.2008526 2.741507 0.3969828
    -1.649275 -0.378958 0.2912694 -0.7183975 1.797741 1.656259
    -0.004479113 -0.6125564
          x29         x30  fitness pset
52 -0.9448366 -0.6126122 68.38073   52
```

This tutorial example shows an improvement of the **sphere** value from 146.3 (first iteration) to 68.4 (last iteration).

5.6 Estimation of Distribution Algorithm

Estimation of Distribution Algorithms (EDA) (Larrañaga and Lozano, 2002) are optimization methods that combine ideas from evolutionary computation, machine learning, and statistics. These methods were proposed in the mid-90s, under several variants, such as Population Based Incremental Learning (PBIL) (Baluja, 1994) and Univariate Marginal Distribution Algorithm (UMDA) (Mühlenbein, 1997).

EDA works by iteratively estimating and sampling a probability distribution that is built from promising solutions (Gonzalez-Fernandez and Soto, 2014). Other population based methods (e.g., evolutionary algorithms) create new individuals using an implicit distribution function (e.g., due to mutation and crossover operators). In contrast, EDA uses an explicit probability distribution defined by a model class (e.g., normal distribution). One main advantage of EDAs is that the search distribution may encode dependencies between the domain problem parameters, thus performing a more effective search.

The EDAs adopted in this chapter are implemented in the **copulaedas** package. The full implementation details are available at (Gonzalez-Fernandez and Soto, 2014). The generic EDA structure is presented in Algorithm 9. The initial population is often created by using a random seeding method. The results of global optimization methods, such as EDAs, can often be enhanced when combined with a local optimization method. Also, as described in Chap. 1, such local optimization can be useful to repair infeasible solutions (see Sect. 5.8 for an example). Then, the population of solutions are improved in a main cycle, until a termination criterion is met.

Within the main cycle, the *selection* function goal is to choose the most interesting solutions. For instance, *truncation selection* chooses a percentage of the best solutions from current population (P), thus working similarly to a greedy selection. The essential steps of EDA are the estimation and simulation of the search distribution, which is implemented by the *learn* and *sample* functions. The learning estimates the structure and parameters of the probabilistic model (M), and the sampling is used to generate new solutions (P') from the probabilistic model. Finally, the *replacement* function defines the next population. For instance, by replacing the full current population by the newly sampled one (P'), by maintaining only the best solutions (found in both populations), or by keeping a diverse set of solutions.

Algorithm 9 Generic EDA pseudo-code implemented in **copulaedas** package, adapted from (Gonzalez-Fernandez and Soto, 2014)

```
1: Inputs: f, C                 ▷ f is the fitness function, C includes control parameters (e.g., N_P)
2: P ← initialization(C)                      ▷ set initial population (seeding method)
3: if required then P ← local_optimization(P, f, C)    ▷ apply local optimization to P
4: end if
5: B ← best(P, f)                                      ▷ best solution of the population
6: i ← 0                                     ▷ i is the number of iterations of the method
7: while not termination_criteria(P, f, C) do
8:     P' ← selection(P, f, C)                                ▷ selected population P'
9:     M ← learn(P')                 ▷ set probabilistic model M using a learning method
10:    P' ← sample(M)           ▷ set sampled population from M using a sampling method
11:    if required then P' ← local_optimization(P', f, C)  ▷ apply local optimization to P'
12:    end if
13:    B ← best(B, P', f)                              ▷ update best solution (if needed)
14:    P ← replacement(P, P', f, C)  ▷ create new population using a replacement method
15:    i ← i + 1
16: end while
17: Output: B                                                          ▷ best solution
```

The **copulaedas** package implements EDAs based on *copula* functions (Joe, 1997), under a modular object oriented implementation composed of separated generic functions that facilitates the definition of new EDAs (Gonzalez-Fernandez and Soto, 2014). EDA components, such as learning and sampling methods, are independently programmed under a common structure shared by most EDAs. The package uses S4 classes, which denotes R objects that have a formal definition of a class (type `> help("Classes")` for more details) and generic methods that can be defined by using the **setMethod** R function. An S4 instance is composed of slots, which is a class component that can be accessed and changed using the **@** symbol. A S4 class instance can be displayed at the console by using the **show()** R function.

The main function is **edaRun**, which implements Algorithm 9, assumes a minimization goal, and includes four arguments:

- **eda**—an EDA instance;
- **f**—evaluation function to be minimized; and
- **lower**, **upper**—lower and upper bounds.

The result is an **EDAResult** class with several slots, namely: **@eda**—EDA class; **@f**—evaluation function; **@lower** and **@upper**—lower and upper bounds; **@numGens**—number of generations (iterations); **@fEvals**—number of evaluations; **@bestEval**—best evaluation; **@bestSol**—best solution; and **@cpuTime**—time elapsed by the algorithm;

An EDA instance can be created using one of two functions, according to the type of model of search distributions: **CEDA**—using multivariate copula; and **VEDA**—using *vines* (graphical models that represent high-dimensional distributions and that can model a more rich variety of dependencies). The main arguments of **CEDA** are: **copula**—**"indep"** (independence or product copula) or **"normal"** (normal copula, the default); **margin**—marginal distribution (e.g., **"norm"**); and **popSize**—population size (N_P, default is 100). The **VEDA** function includes the same **margin** and **popSize** arguments and also: **vine**—**"CVine"** (canonical vine) or **"DVine"** (the default); **copulas**—candidate copulas: **"normal"**, **"t"**, **"clayton"**, **"frank"**, or **"gumbel"** (default is **c("normal")**); and **indepTestSigLevel**—significance independence test level (default **0.01**). The result is a **CEDA** (or **VEDA**) class with two slots: **@name**—the EDA name; and **@parameters**—the EDA parameters. Using these two functions, several EDAs can be defined, including UMDA, Gaussian copula EDA (GCEDA), C-vine EDA (CVEDA), and D-vine (DVEDA):

```
# four EDA types:
# adapted from (Gonzalez-Fernandez and Soto, 2012)
UMDA=CEDA(copula="indep",margin="norm"); UMDA@name="UMDA"
GCEDA=CEDA(copula="normal",margin="norm"); GCEDA@name="GCEDA"
CVEDA=VEDA(vine="CVine",indepTestSigLevel=0.01,
           copulas = c("normal"),margin = "norm")
CVEDA@name="CVEDA"
DVEDA=VEDA(vine="DVine",indepTestSigLevel=0.01,
           copulas = c("normal"),margin = "norm")
DVEDA@name="DVEDA"
```

The population size (N_P) is a critical factor of EDA performance, if too small then the estimate of the search distributions might be inaccurate, while a too large number increases the computational effort and might not introduce any gain in the optimization. Thus, several population size values should be tested. In particular, the **copulaedas** package presents a bisection method that starts with an initial interval and that is implemented using the **edaCriticalPopSize** function (check `> ?edaCriticalPopSize`).

The **copulaedas** package includes several other generic methods that can be defined using the **setMethod** function (type `> help("EDA-class")` for more details), such as: **edaSeed**—initialization function (default is **edaSeedUniform**); **edaOptimize**—local optimization (disabled by default, Sect. 5.8 exemplifies how to define a different function); **edaSelect**—selection function (default is **edaSelectTruncation**); **edaReplace**—replacement function (default is **edaReplaceComplete**—P is replaced by P'); **edaReport**—reporting function (disabled by default); and **edaTerminate**—termination criteria (default **edaTerminateMaxGen**—maximum of iterations).

The same **sphere** ($D = 2$) task is used to demonstrate the EDA:

```
### sphere-EDA.R file ###
library(copulaedas)
```

```
sphere=function(x) sum(x^2)

D=2; maxit=10; LP=5
set.seed(12345) # set for replicability

# set termination criterion and report method:
setMethod("edaTerminate","EDA",edaTerminateMaxGen)
setMethod("edaReport","EDA",edaReportSimple)

# set EDA type:
UMDA=CEDA(copula="indep",margin="norm",popSize=LP,maxGen=maxit)
UMDA@name="UMDA (LP=5)"
# run the algorithm:
E=edaRun(UMDA,sphere,rep(-5.2,D),rep(5.2,D))
# show result:
show(E)
cat("best:",E@bestSol,"f:",E@bestEval,"\n")

# second EDA execution, using LP=100:
maxit=10; LP=100;
UMDA=CEDA(copula="indep",margin="norm",popSize=LP,maxGen=maxit)
UMDA@name="UMDA (LP=100)"
setMethod("edaReport","EDA",edaReportDumpPop) # pop_*.txt files
E=edaRun(UMDA,sphere,rep(-5.2,D),rep(5.2,D))
show(E)
cat("best:",E@bestSol,"f:",E@bestEval,"\n")

# read dumped files and create a plot:
pdf("eda1.pdf",width=7,height=7)
j=1; # j-th parameter
i=1;d=read.table(paste("pop_",i,".txt",sep=""))
plot(xlim=c(1,maxit),rep(1,LP),d[,j],pch=19,
     xlab="iterations",ylab=paste("s_",j," value",sep=""))
for(i in 2:maxit)
{ d=read.table(paste("pop_",i,".txt",sep=""))
  points(rep(i,LP),d[,j],pch=19)
}
dev.off()
```

In this example, the UMDA EDA type was selected using two different population sizes: $N_P = 5$ and $N_P = 100$ (the default **copulaedas** value). For the last EDA, the **edaReportDumpPop** report type is adopted, which dumps each population into a different text file (e.g., the first population is stored at **pop_1.txt**). After showing the second EDA result, the dumped files are read using the **read.table** command, in order to create the plot of Fig. 5.8. The execution result of such demonstration code is:

```
> source("sphere-EDA.R")

  Generation        Minimum           Mean      Std. Dev.
           1 7.376173e+00 1.823098e+01 6.958909e+00
           2 7.583753e+00 1.230911e+01 4.032899e+00
           3 8.001074e+00 9.506158e+00 9.969029e-01
```

```
             4 7.118887e+00 8.358575e+00 9.419817e-01
             5 7.075184e+00 7.622604e+00 3.998974e-01
             6 7.140877e+00 7.321902e+00 1.257652e-01
             7 7.070203e+00 7.222189e+00 1.176669e-01
             8 7.018386e+00 7.089300e+00 4.450968e-02
             9 6.935975e+00 7.010147e+00 7.216829e-02
            10 6.927741e+00 6.946876e+00 1.160758e-02

Results for UMDA (LP=5)
Best function evaluation     6.927741e+00
No. of generations           10
No. of function evaluations 50
CPU time                     0.103 seconds
best: 1.804887 -1.915757 f: 6.927741

Results for UMDA (LP=100)
Best function evaluation     5.359326e-08
No. of generations           10
No. of function evaluations 1000
CPU time                     0.036 seconds
best: -0.00013545 0.0001877407 f: 5.359326e-08
```

When only 5 individuals are used, the algorithm only performs a slight optimization (from $f = 7.38$ to $f = 6.93$). However, when a higher population size is adopted ($N_P = 100$), the EDA performs a very good optimization, achieving a value of 5.36×10^{-8} in only 10 generations. Figure 5.8 shows the respective evolution of the first parameter population values, showing a fast convergence toward the optimum zero value.

5.7 Comparison of Population Based Methods

The goal of this section is to compare all previously presented population based algorithms on two tasks (**rastrigin**, $D = 20$; and **bag prices**, $D = 5$). Four continuous optimization methods are compared: evolutionary algorithm, differential evolution, particle swarm optimization (SPSO 2007), and EDA (GCEDA variant). For the second task, each solution is rounded to the nearest integer value (within [$1,$1000]) before computing the profit. Each method is run fifty times for each task. To simplify the analysis, the comparison is made only in terms of aggregated results over the runs (average or percentage of successes) and no confidence intervals or statistical tests are used (check Sects. 2.2, 4.5, and 5.8 for R code examples of more robust statistical comparative analysis).

Similarly to what is discussed in Sect. 4.5, rather than executing a complete and robust comparison, the intention is more to show how population based algorithms can be compared. To provide a fair comparison and to simplify the experimentation setup, the default parameters of the methods are adopted, except for the population size, which is kept the same for all methods ($N_P = 100$ for **rastrigin** and $LP = 50$

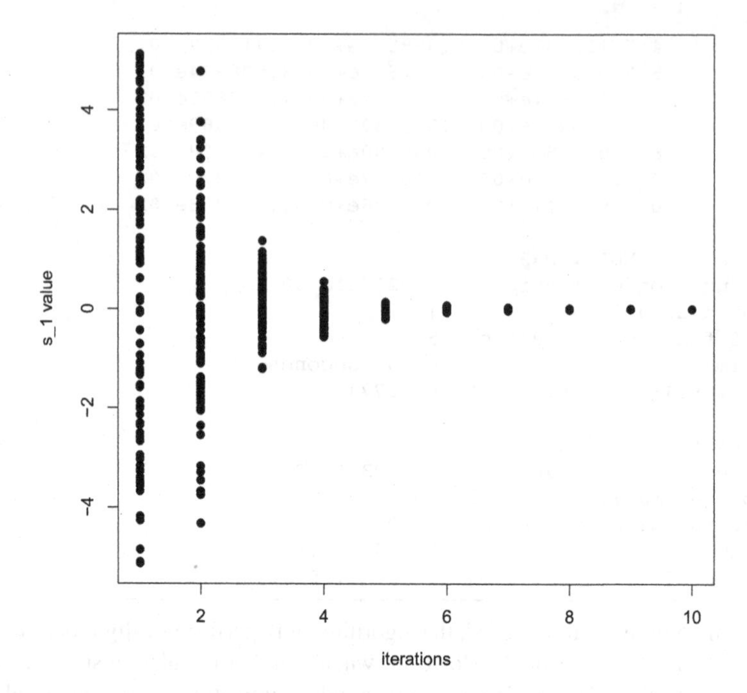

Fig. 5.8 Evolution of the first parameter population values (x_1) for EDA ($N_P = 100$)

for **bag prices**). Also, as performed in Sect. 4.5, all methods are evaluated by storing the best value as a function of the number of evaluations and up to the same maximum number (**MAXFN=10000** for **rastrigin** and **MAXFN=5000** for **bag prices**).

The comparison code (file **compare2.R**) uses the same global variables of Sect. 4.5 (**EV**, **F**, and **BEST**), to store the best value:

```
### compare2.R file ###

source("functions.R") # bag prices functions
library(genalg)
library(DEoptim)
library(pso)
library(copulaedas)

# evaluation functions: ------------------------------------
crastrigin=function(x) # adapted rastrigin
{ f=10*length(x)+sum(x^2-10*cos(2*pi*x))
  # global assignment code: <<-
  EV<<-EV+1 # increase evaluations
  if(f<BEST) BEST<<-f # minimum value
  if(EV<=MAXFN) F[EV]<<-BEST
  return(f)
}
```

```
cprofit=function(x) # adapted bag prices
{ x=round(x,digits=0) # convert x into integer
  # given that EDA occasionally produces unbounded values:
  x=ifelse(x<1,1,x)         # assure that x is within
  x=ifelse(x>1000,1000,x)  # the [1,1000] bounds
  s=sales(x)            # get the expected sales
  c=cost(s)             # get the expected cost
  profit=sum(s*x-c)     # compute the profit
  EV<<-EV+1 # increase evaluations
  if(profit>BEST) BEST<<-profit # maximum value
  if(EV<=MAXFN) F[EV]<<-BEST
  return(-profit) # minimization task!
}
# auxiliary functions: ------------------------------------
crun=function(method,f,lower,upper,LP,maxit) # run a method
{ if(method=="EA")
     rbga(evalFunc=f,stringMin=lower,stringMax=upper,popSize=LP,
          iters=maxit*1.5)
  else if(method=="DE")
     { C=DEoptim.control(itermax=maxit,trace=FALSE,NP=LP)
       DEoptim(f,lower=lower,upper=upper,control=C)
     }
  else if(method=="PSO")
     { C=list(maxit=maxit,s=LP)
       pso::psoptim(rep(NA,length(lower)),fn=f,
               lower=lower,upper=upper,control=C)
     }
  else if(method=="EDA")
     { setMethod("edaTerminate","EDA",edaTerminateMaxGen)
       GCEDA=CEDA(copula="normal",margin="norm",popSize=LP,
                  maxGen=maxit)
       GCEDA@name="GCEDA"
       edaRun(GCEDA,f,lower,upper)
     }
}

successes=function(x,LIM,type="min") # number of successes
{ if(type=="min") return(sum(x<LIM)) else return(sum(x>LIM)) }

ctest=function(Methods,f,lower,upper,type="min",Runs, # test
               D,MAXFN,maxit,LP,pdf,main,LIM) # all methods:
{ RES=vector("list",length(Methods)) # all results
  VAL=matrix(nrow=Runs,ncol=length(Methods)) # best values
  for(m in 1:length(Methods)) # initialize RES object
   RES[[m]]=matrix(nrow=MAXFN,ncol=Runs)

  for(R in 1:Runs) # cycle all runs
    for(m in 1:length(Methods))
      { EV<<-0; F<<-rep(NA,MAXFN) # reset EV and F
        if(type=="min") BEST<<-Inf else BEST<<- -Inf # reset BEST
        suppressWarnings(crun(Methods[m],f,lower,upper,LP,maxit))
        RES[[m]][,R]=F # store all best values
        VAL[R,m]=F[MAXFN] # store best value at MAXFN
      }
```

```
  # compute average F result per method:
  AV=matrix(nrow=MAXFN,ncol=length(Methods))
  for(m in 1:length(Methods))
    for(i in 1:MAXFN)
      AV[i,m]=mean(RES[[m]][i,])
  # show results:
  cat(main,"\n",Methods,"\n")
  cat(round(apply(VAL,2,mean),digits=0)," (average best)\n")
  cat(round(100*apply(VAL,2,successes,LIM,type)/Runs,
           digits=0)," (%successes)\n")

  # create pdf file:
  pdf(paste(pdf,".pdf",sep=""),width=5,height=5,paper="special")
  par(mar=c(4.0,4.0,1.8,0.6)) # reduce default plot margin
  MIN=min(AV);MAX=max(AV)
  # use a grid to improve clarity:
  g1=seq(1,MAXFN,length.out=500) # grid for lines
  plot(g1,AV[g1,1],ylim=c(MIN,MAX),type="l",lwd=2,main=main,
       ylab="average best",xlab="number of evaluations")
  for(i in 2:length(Methods)) lines(g1,AV[g1,i],lwd=2,lty=i)
  if(type=="min") position="topright" else position="bottomright"
  legend(position,legend=Methods,lwd=2,lty=1:length(Methods))
  dev.off() # close the PDF device
}

# define EV, BEST and F:
MAXFN=10000
EV=0;BEST=Inf;F=rep(NA,MAXFN)
# define method labels:
Methods=c("EA","DE","PSO","EDA")
# rastrigin comparison: -----------------------------------
Runs=50; D=20; LP=100; maxit=100
lower=rep(-5.2,D);upper=rep(5.2,D)
ctest(Methods,crastrigin,lower,upper,"min",Runs,D,MAXFN,maxit,LP,
      "comp-rastrigin2","rastrigin (D=20)",75)
# bag prices comparison: ---------------------------------
MAXFN=5000
F=rep(NA,MAXFN)
Runs=50; D=5; LP=50; maxit=100
lower=rep(1,D);upper=rep(1000,D)
ctest(Methods,cprofit,lower,upper,"max",Runs,D,MAXFN,maxit,LP,
      "comp-bagprices","bag prices (D=5)",43500)
```

Two important auxiliary functions were defined: **crun**—for executing a run of one of the four methods; and **ctest**—for executing several runs and showing the overall results. For the evolutionary algorithm, the maximum number of iterations is increased by 50% to assure that at least **MAXFN** evaluations are executed (due to elitism, the number of tested solutions is lower than $N_P \times maxit$). The obtained results for each task are presented in terms of a plot and two console metrics. Each plot shows in the y-axis the evolution of the average best value, while the $x-$axis contains the number of evaluation functions. The two metrics used are: the average (over all runs), the best result (measured at **MAXFN**), and the percentage of successes. The last metric is measured as the proportion of best results below 75 (for **rastrigin**) or above 43500 (for **bag prices**).

The result of the two plots is presented in Fig. 5.9, while the console results are:

```
> source("compare2.R")
rastrigin (D=20)
 EA DE PSO EDA
18 67 98 77   (average best)
100 92 2 42   (%successes)
bag prices (D=5)
 EA DE PSO EDA
43758 43834 43708 43659   (average best)
100 100 100 94   (%successes)
```

The comparison of methods using the methodology related with Fig. 5.9 is interesting, as it presents the average behavior of the method throughout the number of evaluations, which is correlated with computational effort. For the **rastrigin** task, the particle swarm and EDA methods show a faster initial convergence when compared with other methods (see Fig. 5.9) and thus they provide better results if few computational resources are available. However, after a while the particle swarm and EDA convergence becomes more slower and the methods are outperformed by the evolutionary algorithm. Hence, if more computation power is available, then evolutionary algorithm is the best method. At `MAXFN` evaluations, the average best for the evolutionary algorithm is 18, which is substantially lower than other methods. A similar effect occurs with the **bag prices** task, although the optimization curves are much closer. Initially, for a smaller number of evaluations, the EDA and particle swarm methods provide the best results. Yet, after around 2000 evaluations, all methods tend to converge to similar results. At the end of the optimization, the best results are provided by the differential evolution, with an average value of 43834 and 100% of successes. It should be noted that this is just a demonstrative comparison example and different results could be achieved with a distinct experimental setup (e.g., different N_P and `MAXFN` values).

5.8 Bag Prices with Constraint

This section compares two strategies for handling constraints: death-penalty and repair. As explained in Sect. 1.5, death-penalty is a simple strategy that can be easily applied to any optimization method. It requires only changing the evaluation function to return very high value if the solution is infeasible. However, such strategy is not very efficient, since the infeasible solutions do not guide the search, thus behaving similarly to Monte Carlo random search, see Sect. 3.4. The repair alternative tends to be more efficient, as it transforms an infeasible solution into a feasible one, but often requires domain knowledge.

For this comparison, the **bag prices** task (of Sect. 1.8) is adopted with a hard constraint: the maximum number of bags that can be manufactured within a

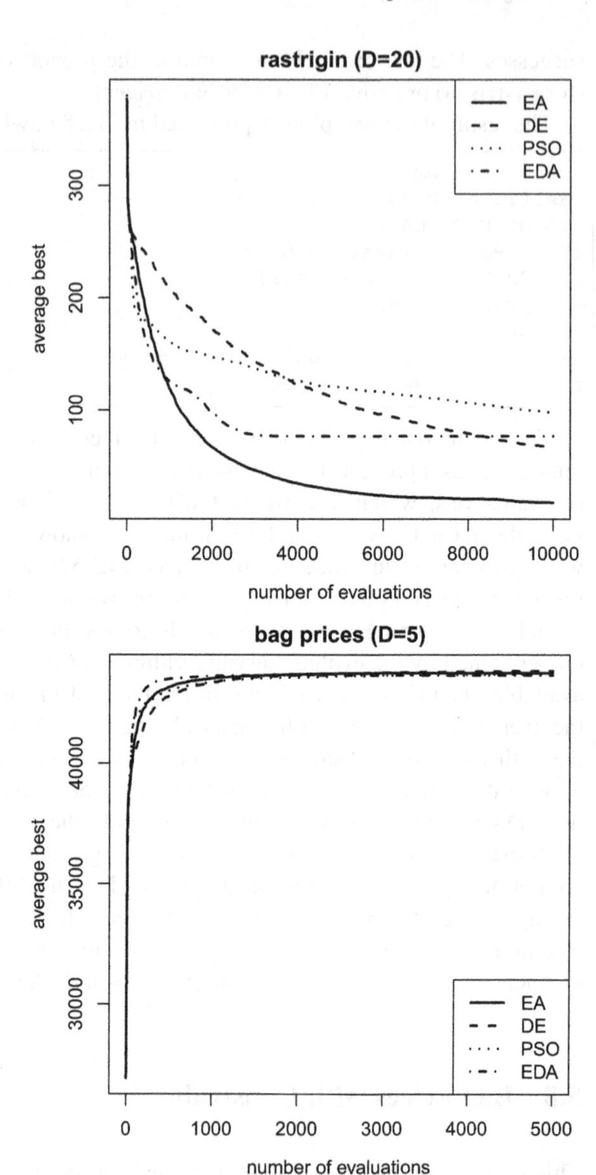

Fig. 5.9 Population based search comparison example for the **rastrigin** (top) and **bag prices** (bottom) tasks

production cycle is set to 50. Also, the EDA method is used as the optimization engine, since the **copulaedas** package presents a useful feature for the repair strategy, since it is possible to add a local optimization method within the EDA main cycle (by using the **edaOptimize** generic method). The death-penalty is simply implemented by returning **Inf** when a solution is infeasible, while the repair method uses a local search and domain knowledge. The code of Sect. 5.7 was adapted (e.g., same N_P, $maxit$, and **MAXFN** values) for this experiment:

```
### bag-prices-constr.R file ###
```

```
source("functions.R") # bag prices functions
library(copulaedas) # EDA

# evaluation function: ------------------------------------
cprofit2=function(x) # bag prices with death-penalty
{ x=round(x,digits=0) # convert x into integer
  x=ifelse(x<1,1,x)          # assure that x is within
  x=ifelse(x>1000,1000,x)  # the [1,1000] bounds
  s=sales(x)
  if(sum(s)>50) res=Inf # if needed, death-penalty!!!
  else{ c=cost(s);profit=sum(s*x-c)
        # if needed, store best value
        if(profit>BEST) { BEST<<-profit; B<<-x}
        res=-profit # minimization task!
      }
  EV<<-EV+1 # increase evaluations
  if(EV<=MAXFN) F[EV]<<-BEST
  return(res)
}
# example of a local search method that repairs a solution:
localRepair=function(eda, gen, pop, popEval, f, lower, upper)
{
 for(i in 1:nrow(pop))
 { x=pop[i,]
   x=round(x,digits=0) # convert x into integer
   x=ifelse(x<lower[1],lower[1],x) # assure x within
   x=ifelse(x>upper[1],upper[1],x) # bounds
   s=sales(x)
   if(sum(s)>50)
   {
    x1=x
    while(sum(s)>50) # new constraint: repair
    { # increase price to reduce sales:
      x1=x1+abs(round(rnorm(D,mean=0,sd=5)))
      x1=ifelse(x1>upper[1],upper[1],x1) # bound if needed
      s=sales(x1)
    }
    x=x1 # update the new x
   }
   pop[i,]=x;popEval[i]=f(x)
 }
 return(list(pop=pop,popEval=popEval))
}

# experiment: ---------------------------------------------
MAXFN=5000
Runs=50; D=5; LP=50; maxit=100
lower=rep(1,D);upper=rep(1000,D)
Methods=c("Death","Repair")
setMethod("edaTerminate","EDA",edaTerminateMaxGen)
GCEDA=CEDA(copula="normal",margin="norm",popSize=LP,
           maxGen=maxit,fEvalStdDev=10)
GCEDA@name="GCEDA"
```

```
RES=vector("list",length(Methods)) # all results
VAL=matrix(nrow=Runs,ncol=length(Methods)) # best values
for(m in 1:length(Methods)) # initialize RES object
   RES[[m]]=matrix(nrow=MAXFN,ncol=Runs)
for(R in 1:Runs) # cycle all runs
  {
    B=NA;EV=0; F=rep(NA,MAXFN); BEST= -Inf # reset vars.
    setMethod("edaOptimize","EDA",edaOptimizeDisabled)
    setMethod("edaTerminate","EDA",edaTerminateMaxGen)
    suppressWarnings(edaRun(GCEDA,cprofit2,lower,upper))
    RES[[1]][,R]=F # store all best values
    VAL[R,1]=F[MAXFN] # store best value at MAXFN

    B=NA;EV=0; F=rep(NA,MAXFN); BEST= -Inf # reset vars.
    # set local repair search method:
    setMethod("edaOptimize","EDA",localRepair)
    # set additional termination criterion:
    setMethod("edaTerminate","EDA",
              edaTerminateCombined(edaTerminateMaxGen,
                  edaTerminateEvalStdDev))
    # this edaRun might produces warnings or errors:
    suppressWarnings(try(edaRun(GCEDA,cprofit2,lower,upper),
        silent=TRUE))
    if(EV<MAXFN) # if stopped due to EvalStdDev
       F[(EV+1):MAXFN]=rep(F[EV],MAXFN-EV) # replace NAs
    RES[[2]][,R]=F # store all best values
    VAL[R,2]=F[MAXFN] # store best value at MAXFN
  }

# compute average F result per method:
AV=matrix(nrow=MAXFN,ncol=length(Methods))
for(m in 1:length(Methods))
  for(i in 1:MAXFN)
    AV[i,m]=mean(RES[[m]][i,])
# show results:
cat(Methods,"\n")
cat(round(apply(VAL,2,mean),digits=0)," (average best)\n")
# Mann-Whitney non-parametric test:
p=wilcox.test(VAL[,1],VAL[,2],paired=TRUE)$p.value
cat("p-value:",round(p,digits=2),"(<0.05)\n")

# create pdf file:
pdf("comp-bagprices-constr.pdf",width=5,height=5,paper="special")
par(mar=c(4.0,4.0,1.8,0.6)) # reduce default plot margin
# use a grid to improve clarity:
g1=seq(1,MAXFN,length.out=500) # grid for lines
plot(g1,AV[g1,2],type="l",lwd=2,
     main="bag prices with constraint",
     ylab="average best",xlab="number of evaluations")
lines(g1,AV[g1,1],lwd=2,lty=2)
legend("bottomright",legend=rev(Methods),lwd=2,lty=1:4)
dev.off() # close the PDF device
```

The two constraint handling methods are compared similarly as in Sect. 5.7, thus global **EV**, **F**, and **BEST** variables are used to store the best values at a given function evaluation. The death-penalty is implemented in function **cprofit2**. The repair solution is handled by the **localRepair** function, which contains the signature (function arguments) required by **edaOptimize** (e.g., eda argument is not needed). If a solution is infeasible, then a local search is applied, where the solution prices are randomly increased until the expected sales is lower than 51. This local search uses the following domain knowledge: the effect of increasing a price is a reduction in the number of sales. The new feasible solutions are then evaluated. To reduce the number of code lines, the same **cprofit2** function is used, although the death-penalty is never applied, given that the evaluated solution is feasible. The **localRepair** function returns a list with the new population and evaluation values. Such list is used by **edaRun** to replace the sampled population (P'), thus behaving as a Lamarckian global-local search hybrid (see Sect. 1.6).

The EDA applied is similar to the one presented in Sect. 5.7 (e.g., GCEDA), except for the repair strategy, which includes the explained local search and an additional termination criterion (**edaTerminateEvalStdDev**), which stops when the standard deviation of the evaluation of the solutions is too low. When more than one criterion is used in EDA, the **edaTerminateCombined** method needs to be adopted. The **edaTerminateEvalStdDev** extra criterion was added given that the repair strategy leads to a very fast convergence and thus the population quickly converges to the same solution. However, setting the right standard deviation threshold value (**fEvalStdDev**) is not an easy task (setting a too large value will stop the method too soon). With **fEvalStdDev=10**, **edaRun** still occasionally produces warnings or errors. Thus, the **suppressWarnings** and **try** R functions are used to avoid this problem. The latter function prevents the failure of the **edaRun** execution. Such failure is not problematic, given that the results are stored in global variables. In the evaluation function (**cprofit2**), the global variable **B** is used to store the best solution, which is useful when **edaRun** fails. When a failure occurs, the code also replaces the remaining **NA** values of the **F** object with the last known evaluation value.

The obtained results are processed similarly to what is described in Sect. 5.7, except that the number of successes is not measured and the final average results are compared with a Mann–Whitney nonparametric statistical test (using the **wilcox.test** function). An example of file **bag-prices-constr.R** execution is:

```
> source("bag-prices-constr.R")
Death Repair
31028 32351   (average best)
p-value: 0 (<0.05)
```

As shown by the average results, statistical test (p-value<0.05), and plot of Fig. 5.10, the repair strategy clearly outperforms the death-penalty one, with a final average difference of $1323 (statistically significant). Nevertheless, it should be

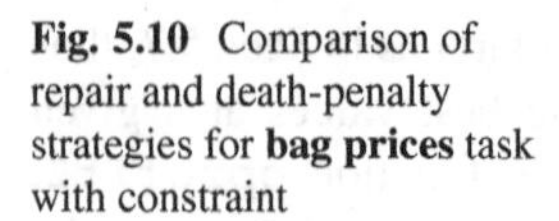
Fig. 5.10 Comparison of repair and death-penalty strategies for **bag prices** task with constraint

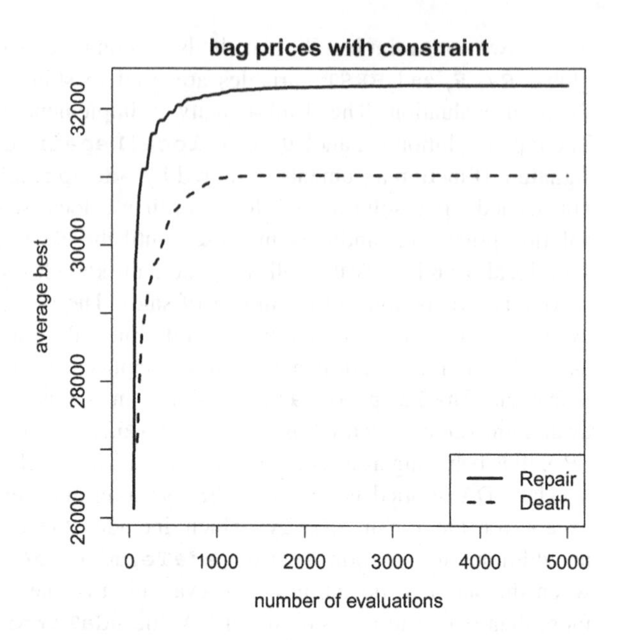

noted that this repair strategy uses domain knowledge and thus it cannot be applied directly to other tasks.

5.9 Parallel Execution of Population Based Methods

As explained in Sect. 2.5, the computational execution time can be reduced by splitting a computer program into smaller pieces that are run using distinct processors or cores. When population based methods are used, each individual evaluation is an independent process, since the returned quality is computed using only the solution value. Therefore, population based methods can easily be adopted to a parallel execution by running each solution evaluation in a different processing machine. There are also more sophisticated parallel approaches. One example is the island model, which is applicable to a wide range of evolutionary computation algorithms (Scrucca, 2017; Hashimoto et al., 2018). The model assumes that there are several islands, each containing a sub-population of solutions that are evolved and that can be executed on a distinct processor. On a regular basis, assuming a migration iteration interval, the worst solutions in an island are replaced by the best solutions from neighbor islands. The simpler and island parallel distribution schemes are depicted in Fig. 5.11.

Before presenting the demonstration code, two relevant aspects are highlighted. First, the usage of parallel computation can imply system and data memory communication computational costs. Thus, the parallel evaluation might only

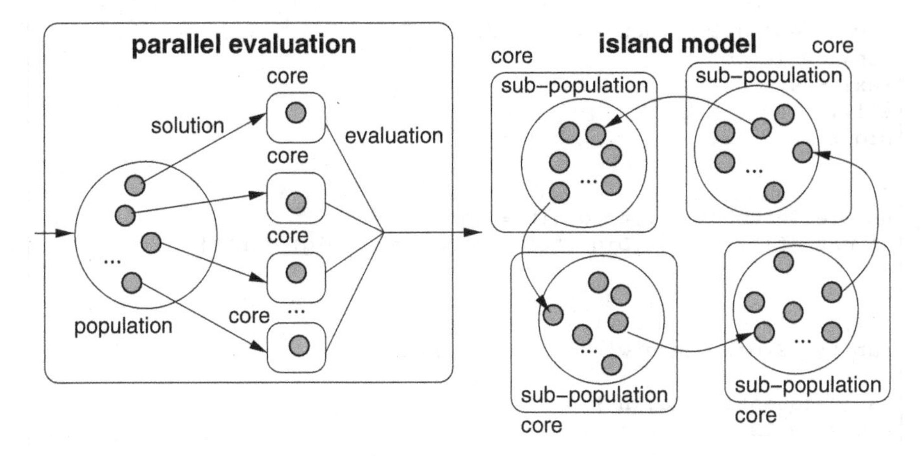

Fig. 5.11 Example of parallel evaluation (left) and the island distribution (right) models

provide reduced computational times when compared with the sequential evaluation if the evaluation function is computationally demanding. Second, some parallel approaches, such as the island model, change the way the search is performed, thus the optimized value can be different. In effect, the parallel evaluation scheme will always produce the same optimization value as provided by the sequential execution. However, this is not the case of the island model, which works differently from the standard evolutionary algorithm. The usage of different sub-populations increases diversity, while the migration of solutions prevents each island from converging to a local minimum or maximum (Hashimoto et al., 2018).

The demonstration uses the **rastrigin** task (with $D = 30$). The first example code (**`rastrigin-p1.R`**) compares the sequential and simple parallel executions for two population based algorithms, evolutionary algorithms (**`GA`** package) and differential evolution (**`DEoptim`**):

```
### rastrigin-p1.R file ###

# packages with parallel execution:
library(GA)       # load ga
library(DEoptim) # load DEoptim

# package to measure time elapsed:
library(tictoc)     # load tic and toc

# computationally slow rastrigin function:
srastrigin=function(x)
{ res=10*length(x)+sum(x^2-10*cos(2*pi*x))
  Sys.sleep(0.1) # sleep ms
  return(res)
}
srastrigin2=function(x) -srastrigin(x) # maximization goal
D=30 # dimension

# common setup for the executions:
```

```
lower=rep(-5.2,D);upper=rep(5.2,D) # lower and upper bounds
Pop=100 # population size
Maxit=20 # maximum number of iterations
Eli=1 # elitism of one individual
Dig=2 # use 2 digits to show results

# show best solution, evaluation value and time elapsed:
showres2=function(f,tm,digits=Dig)
 { cat("f:",round(f,Dig),"time:",round(tm,Dig),"\n")}

# Number of processing Cores (change if needed):
NC=4 # NC=detectCores() is another option
cat(">> Experiments with:",NC,"cores\n")

cat("sequential ga:\n")
set.seed(123) # set for replicability
tic()
ga1=ga(type="real-valued",fitness=srastrigin2,
       lower=lower,upper=upper, # lower and upper bounds
       popSize=Pop, # population size
       elitism=Eli,maxiter=Maxit,monitor=FALSE)
te=toc(quiet=TRUE) # time elapsed
showres2(-ga1@fitnessValue,te$toc-te$tic)

cat("parallel ga:\n")
set.seed(123) # set for replicability
tic()
ga2=ga(type="real-valued",fitness=srastrigin2,
       lower=lower,upper=upper, # lower and upper bounds
       popSize=Pop,elitism=Eli,maxiter=Maxit,
       parallel=NC, # NC cores for parallel evaluation
       monitor=FALSE)
te=toc(quiet=TRUE) # time elapsed
showres2(-ga2@fitnessValue,te$toc-te$tic)

cat("sequential DEoptim:\n") # minimization
set.seed(123) # set for replicability
tic()
C=DEoptim.control(NP=Pop,itermax=Maxit,trace=FALSE)
de1=suppressWarnings(DEoptim(srastrigin,lower,upper,control=C))
te=toc(quiet=TRUE) # time elapsed
showres2(de1$optim$bestval,te$toc-te$tic)

cat("parallel DEoptim:\n") # minimization
set.seed(123) # set for replicability
cl=makeCluster(NC) # set NC cores
tic()
C=DEoptim.control(NP=Pop,itermax=Maxit,trace=FALSE,parallelType
    =1,cluster=cl)
de2=suppressWarnings(DEoptim(srastrigin,lower,upper,control=C))
te=toc(quiet=TRUE) # time elapsed
stopCluster(cl) # stop the cluster
showres2(de2$optim$bestval,te$toc-te$tic)
```

To show the parallel benefits, a new **srastrigin** function is defined, which corresponds to a "slower" **rastrigin** that requires at least 10 ms of execution time (due to the usage of the **System.sleep** R command, explained Sect. 2.5). The running setup assumes **NC=4** cores, usage of the same random initialization seed for each algorithm (via **set.seed**), a population of **Pop=100** individuals, and a total of **Maxit=20** iterations. The default operator values are adopted for the **GA** package, which includes (Michalewicz, 1996): a linear rank selection; a local arithmetic crossover—which linearly combines two parents; and a uniform random mutation. Also, the **tic** and **toc** functions (Sect. 2.5) are used to measure the time elapsed, in seconds. After running the algorithm, the code shows the best optimized value and total time elapsed. For the differential evolution, the **makeCluster** function (Sect. 2.5) needs to be explicitly called, since the **parallelType=1** argument of **DEoptim.control()** assumes a default usage of all cores (the machine where the code was run has a total of 8 cores). The result of file **rastrigin-p1.R** is:

```
> source("rastrigin-p1.R")
>> Experiments with: 4 cores
sequential ga:
f: 202.93 time: 170.71
parallel ga:
f: 202.93 time: 44.75
sequential DEoptim:
f: 306.06 time: 212.74
parallel DEoptim:
f: 306.06 time: 54.84
```

In this case, the parallel execution of the generic algorithm is around 3.81 times faster than its sequential version, requiring just 44.75 s and optimizing the same best solution (evaluation value of 202.93). Similarly, the parallel differential evolution execution is around 3.95 times faster than the sequential variant, obtaining an evaluation value of 306.06 in 54.84 s. Overall, the parallel speedup is very close to the number of cores used (**NC=4**).

The same **rastrigin** task (with $D = 30$) is used for the island model demonstration, which is implemented in the **GA** package in terms of the **gaisl** function. The demonstration file **rastrigin-p2.R** uses the standard **rastrigin** function, since the slower version (**srastrigin**) produces a very large **gaisl()** execution time:

```
### rastrigin-p2.R file ###

library(GA)      # load ga and gaisl
library(tictoc) # load tic and toc

# real value rastrigin function with a dimension of 30:
rastrigin=function(x) 10*length(x)+sum(x^2-10*cos(2*pi*x))
rastrigin2=function(x) -rastrigin(x) # maximization goal
D=30

# common setup for the executions:
```

```
lower=rep(-5.2,D);upper=rep(5.2,D) # lower and upper bounds
Pop=1000 # population size
Maxit=20 # maximum number of iterations
Dig=2 # use 2 digits to show results
Migrate=5 # migrate solutions every 5 iterations

# show best solution, evaluation value and time elapsed:
showres2=function(f,tm,digits=Dig)
 { cat("f:",round(f,Dig),"time:",round(tm,Dig),"\n")}

# Number of processing Cores (change if needed):
NC=detectCores()
cat(">> Experiments with:",NC,"cores\n")

cat("sequential ga:\n")
set.seed(123) # set for replicability
tic()
ga1=ga(type="real-valued",fitness=rastrigin2,
       lower=lower,upper=upper, # lower and upper bounds
       popSize=Pop, # population size
       maxiter=Maxit,monitor=FALSE)
te=toc(quiet=TRUE) # time elapsed
showres2(-ga1@fitnessValue,te$toc-te$tic)

cat("sequential island:\n")
set.seed(123) # set for replicability
tic()
ga2=gaisl(type="real-valued",fitness=rastrigin2,
       lower=lower,upper=upper, # lower and upper bounds
       popSize=Pop,
       numIslands=NC, # number of Islands
       migrationInterval=Migrate,
       parallel=FALSE, # do not use
       monitor=FALSE)
te=toc(quiet=TRUE) # time elapsed
showres2(-ga2@fitnessValue,te$toc-te$tic)

cat("islands ga (GA package):\n")
set.seed(123) # set for replicability
tic()
ga3=gaisl(type="real-valued",fitness=rastrigin2,
       lower=lower,upper=upper, # lower and upper bounds
       popSize=Pop,
       numIslands=NC, # number of Islands
       migrationInterval=Migrate,
       parallel=TRUE, # use the cores
       monitor=FALSE)
te=toc(quiet=TRUE) # time elapsed
showres2(-ga3@fitnessValue,te$toc-te$tic)
```

The code is similar to the previous demonstration, although it includes a few changes. The number of used cores is now obtained using the automatic **detectCores()** R function. Also, the population size was increased to **Pop=1000** and the default elitism values are adopted (**gaisl** uses a default elitism value of top 5% individuals in each island). Moreover, the code assumes three main algorithm executions: a sequential genetic algorithm, a sequential island model, and a parallel island run. Both island models use **NC** islands or sub-populations, the difference is that in the parallel run each sub-population is executing using a different core. In the R console, the obtained result is:

```
>> Experiments with: 8 cores
sequential ga:
f: 74.39 time: 0.46
sequential island:
f: 0 time: 25.38
islands ga (GA package):
f: 0 time: 12.03
```

The tested machine has 8 cores. While only one core was used by the sequential standard generic algorithm, it produced a very fast execution, requiring just 0.46 s. In contrast, the sequential island model, which is more complex and performs several sub-population exchanges (at every **Migrate** iterations), required 25.38 s of processing time but optimized a better solution (in effect, it obtained the perfect solution, with an evaluation of 0.0). As for the parallel island run, it performs the same island search but it is around 2.1 times faster than its sequential version.

5.10 Genetic Programming

Genetic programming denotes a collection of evolutionary computation methods that automatically generate computer programs (Banzhaf et al., 1998). In general, the computer programs have a variable length and are based on lists or trees (Luke, 2015). Hence, the goal of genetic programming is quite distinct from the previous presented population based algorithms. Instead of numerical optimization, genetic programming is used in tasks such as automatic programming or discovering mathematical functions. As pointed out in Flasch (2014), the one key advantage of genetic programming is that the representation of the solutions is often easy to interpret by humans; however, the main drawback is a high computational cost, due to the high search space of potential solutions.

Genetic programming adopts the same concepts of evolutionary algorithms, with a population of solutions that compete for survival and use of genetic operators for generating new offspring. Thus, the search engine is similar to what is described in Algorithm 5. Given that a different representation is adopted (e.g., trees), a distinct initialization function is adopted (e.g., *random growth*) and specialized genetic operators are used (Michalewicz et al., 2006). There are two main mutation types in classical genetic programming systems: replace a randomly selected value or

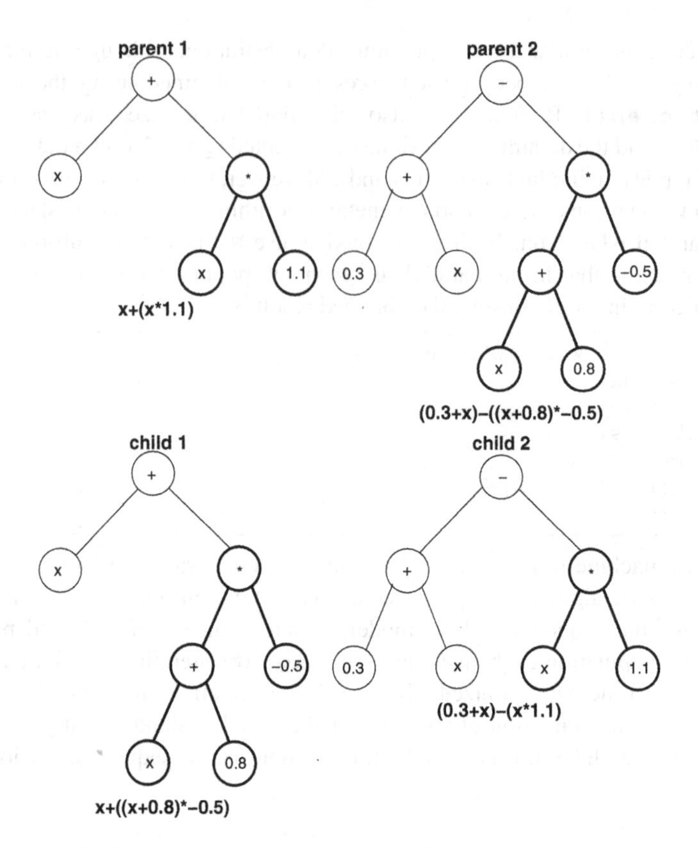

Fig. 5.12 Example of a genetic programming random subtree crossover

function by another random value or function; and replace a randomly selected subtree by another generated subtree. The classical crossover works by replacing a random subtree from a parent solution by another random subtree taken from a second parent. Figure 5.12 shows an example of the random subtree crossover and includes four examples of tree representations for mathematical functions.

The first edition of this book explored the **rgp** package for a mathematical function discovery. Yet, the package is not currently available at the CRAN server for the tested R version (4.0.0), which constitutes an example of a technical debt (Sect. 4.3). The suggested turn around is to install the last **rgp** archived version, following the steps shown in file **install-rgp.R**:

```
### install-rgp.R file ###

install.packages("devtools") # install first devtools
library(devtools) # load devtools
install.packages("emoa") # install emoa rgp dependency
install_version("rgp",version="0.4-1") # install archived rgp
```

Executing this file in R should install the package. The `devtools` package allows the installation of packages that contain compiled code. The `emoa` package also needs to be explicitly installed, since it is used by `rgb`. The last instruction, `install_version` performs the installation of older package versions, in this case `rgp` version `"0.4-1"`. If the suggested code does not work and the package cannot be installed, then the alternative is to explore the grammatical evolution code (Sect. 5.11), which performs a similar approximation function.

The genetic programming demonstration intention is to show how non-numerical representations can be handled by a modern optimization technique. Only a brief explanation of the `rgp` features is provided, since the package includes a large range of functions. If needed, further details can be obtained in Flasch (2014) (`> vignette("rgp_introduction")`).

The first step is to define the symbolic expression search space. In `rgp`, solutions are represented as R functions, which are constructed in terms of three sets: input variables (function arguments), constants, and function symbols. Constants are created in terms of factory functions, which typically are stochastic and are called each time a new constant is needed. The function symbols usually include arithmetic operators, such as addition (`+`) or subtraction (`-`). Other R mathematical functions can also be included (e.g., `exp`, `log`), but some care is needed to avoid invalid expressions (e.g., `log(-1)` returns a `NaN`). These sets (whose members are known as building blocks) can be set using the `rgp` functions:

- `inputVariableSet` —these arguments are the names (strings) of input variables;
- `constantFactorySet`—this argument is a factory function that often includes a random number generator (e.g., `rnorm`, `runif`); and
- `functionSet`—these arguments are strings that define the set of mathematical symbols.

Next, an evaluation function needs to be defined. The `rgp` assumes a minimization goal. The final step consists in selecting the genetic programming parameters and running the algorithm. This is achieved using the `geneticProgramming` function, which includes parameters such as:

- `fitnessFunction`—evaluation function that includes one argument (the expression);
- `stopCondition`—termination criteria (usually based on maximum runtime, in seconds; `?makeStepsStopCondition` shows other stopping options and full details);
- `population`—initial population, if missing it is created using a random growth;
- `populationSize`—size of the population (N_P, the default is 100);
- `eliteSize`—number of individuals to keep (*elitism*, defaults to `ceiling (0.1 * populationSize)`);
- `functionSet`—set of mathematical symbols;
- `inputVariables`—set of input variables;

- **constantSet**—set of constant factory functions;
- **crossoverFunction**—crossover operator (defaults to **crossover**, which is the classical random subtree crossover, see Fig. 5.12);
- **mutationFunction**—mutation function (defaults to **NULL**, check **?mutateFunc** for **rgp** mutation possibilities);
- **progressMonitor**—function called every generation and that shows the progress of the algorithm; and
- **verbose**—if progress should be printed.

The result is a list with several components, including: **$population**, the last population; and **$fitnessValues**, the evaluation values of such population.

To show the **rgp** capabilities, the synthetic **rastrigin** function ($D = 2$) is adopted and approximated with a polynomial function. Thus, the variable set includes two inputs (x_1 and x_2) and the set of function symbols is defined as {'*','+','-'} (3 simple arithmetic operators: multiply, add, and subtract). This setup allows to define a wide range of symbolic expressions, including for instance $x_1 + x_1 \times x_2 - 3.4$ and $-23.4 \times x_2 + 0.32$. In the demonstration code, the set of constants is generated using a normal distribution. Moreover, the evaluation function is not the **rastrigin** function itself, since the goal is to approximate this function. Rather, the evaluation function is set as the error between the rastrigin and candidate expression outputs for an input domain. The Mean Squared Error (MSE) is the adopted error metric. Widely used in statistics, the metric is defined as $MSE = \sum_{i=1}^{N}(y_i - \hat{y}_i)^2/N$, where y_i is the desired target value for input $\mathbf{x}_i$, $\hat{y}_i$ is the estimated value, and N is the number of input examples. MSE penalizes higher individual errors, and the lower the metric, the better is the approximation. The genetic programming is set with a population size of $N_P = 100$, a random subtree mutation (with maximum subtree depth of 4, using the **mutateSubtree rgp** function) and it is stopped after 50 s. The implemented R code is:

```
### rastrigin-gp.R ###

library(rgp) # load rgp

# auxiliary functions:
rastrigin=function(x) 10*length(x)+sum(x^2-10*cos(2*pi*x))
fwrapper=function(x,f) f(x[1],x[2]) # auxiliar function

# configuration of the genetic programming:
VS=inputVariableSet("x1","x2")
cF1=constantFactorySet(function() rnorm(1)) # mean=0, sd=1
FS=functionSet("+","*","-") # simple arithmetic operators
Pop=100
Gen=100
# simple monitor function: show best every 10 iterations:
monitor=function(population,objectiveVectors,fitnessFunction,
                 stepNumber,evaluationNumber,bestFitness,
                 timeElapsed,...)
{ if(stepNumber==2||stepNumber%%10==0)
    cat("iter:",stepNumber,"f:",bestFitness,"\n")
```

```
}

# set the input samples (grid^2 data points):
grid=10 # size of the grid used
domain=matrix(ncol=2,nrow=grid^2) # 2D domain grid
domain[,1]=rep(seq(-5.2,5.2,length.out=grid),each=grid)
domain[,2]=rep(seq(-5.2,5.2,length.out=grid),times=grid)

y=apply(domain,1,rastrigin) # compute target output

msevalue=function(f) # mse evaluation of function f
{ mse(y,apply(domain,1,fwrapper,f)) }

# run the genetic programming:
set.seed(12345) # set for replicability
mut=function(func) # set the mutation function
{ mutateSubtree(func,funcset=FS,inset=VS,conset=cF1,
                mutatesubtreeprob=0.1,maxsubtreedepth=4) }

# call the genetic programming:
gp=geneticProgramming(fitnessFunction=msevalue,
                      populationSize=Pop,
                      stopCondition=makeStepsStopCondition(Gen),
                      functionSet=FS,inputVariables=VS,
                      constantSet=cF1,mutationFunction=mut,
                      progressMonitor=monitor,verbose=TRUE)
# show the results:
b=gp$population[[which.min(gp$fitnessValues)]]
cat("best solution (f=",msevalue(b),"):\n")
print(b)

# create approximation pdf plot:
y2=apply(domain,1,fwrapper,b)
MIN=min(y,y2);MAX=max(y,y2)
pdf("gp-function.pdf",width=7,height=7,paper="special")
plot(y,ylim=c(MIN,MAX),type="l",lwd=2,lty=1,
     xlab="points",ylab="function values")
lines(y2,type="l",lwd=2,lty=2)
legend("bottomright",leg=c("rastrigin","GP function"),lwd=2,
       lty=1:2)
dev.off()
```

In this example, the two input variables are named **"x1"** and **"x2"**, while the input domain is created as a two-dimensional grid, where each input is varied within the range $[-5.2, 5.2]$, with a total of **grid*grid**=100 samples. The **domain** matrix is created using the **rep** and **seq** R functions (type `> print(domain)` to check the matrix values). Such matrix is used by the **msevalue** function, which uses the **apply** function at the row level to generate first the **rastrigin** and expression (**f**) outputs and then computes the MSE for all domain samples. The **mse** function computes the MSE and it is defined in the **rgp** package. The auxiliary **fwrapper** function was created for an easier use of the **apply** instruction over the **f** object, since **f** receives two arguments while a row from **domain** is one vector with two elements. The execution includes a simple monitoring function (**monitor**) that shows the evolution of the best fitness value. It should be noted that the **rgp** implementation only calls this function after generation 2. After running the genetic programming, the best solution is presented. The fittest solution is stored in object **b**, which is a function with two arguments (**x1** and **x2**). Also, the code creates a PDF file, related with a two-dimensional plot, where the $x-$axis denotes all 100 points and the $y-$ axis represents the rastrigin and genetic programming best solution output values. The result of running the demonstration file (**rastrigin-gp.R**) is:

```
> source("rastrigin-gp.R")
STARTING genetic programming evolution run (Age/Fitness/
    Complexity Pareto GP search-heuristic) ...
iter: 2 f: 857.2384
iter: 10 f: 451.9328
iter: 20 f: 433.0521
iter: 30 f: 337.4354
iter: 40 f: 299.4179
iter: 50 f: 229.8855
iter: 60 f: 192.6944
iter: 70 f: 192.6944
iter: 80 f: 192.2054
iter: 90 f: 192.2054
iter: 100 f: 192.2054
evolution step 100, fitness evaluations: 4950, best fitness: 192
    .205414, time elapsed: 8.15\,s
Genetic programming evolution run FINISHED after 100 evolution
    steps, 4950 fitness evaluations and 8.15\,s.
best solution (f= 192.2054 ):
function (x1, x2)
x2 * x2 - -1.85041442021454 - (-1.85041442021454 - (x1 * x1 -
    -1.85041442021454))
<bytecode: 0x7ff8d0220418>
```

After 100 generations, the best obtained solution is $x_1^2 + x_2^2 - 1.85041442021454$. The algorithm evolved a MSE value from 857.2 (generation 2) to 192.2 (generation 100). Figure 5.13 shows the created plot, revealing an interesting fit. It should be noted that the final solution was obtained by performing a manual "cleaning" of the returned symbolic expression. Such post-processing (using manual or automatic techniques) is a common task when human understandable knowledge is required.

5.11 Grammatical Evolution

Grammatical evolution is an evolutionary algorithm proposed by Ryan et al. (1998). Similarly to genetic programming, the method is capable of evolving programs of arbitrary size. The main difference is that the generated programs follow a syntax that is defined through a grammar. Grammars are known in the domain of computer

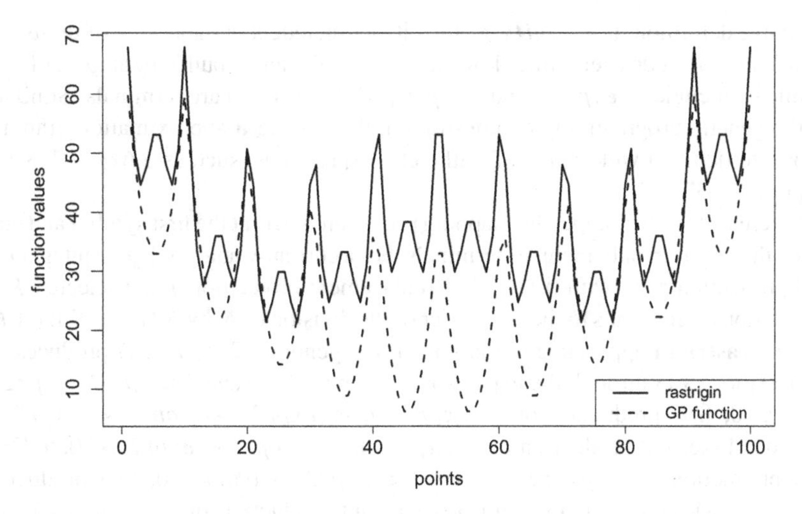

Fig. 5.13 Comparison of **rastrigin** function and best solution given by the genetic programming

science, since they are used to describe the structure of programming languages (e.g., C, R, Python). In grammatical evolution, the genome is composed of an array of integers that are expanded into a program (the phenotype). The initial population is often set randomly and an evolutionary algorithm is used to evolve a population of solutions, using the typical crossover, mutation, and selection operations. When compared with the standard generic programming, grammatical evolution is more flexible, as it can work with any program structure (defined by the grammar) and not just tree structures.

A key element of a grammatical evolution is the grammar definition, which assumes a context-free BackusNaur Form (BNF), containing terminal or non-terminal symbols. Terminal symbols are items of the language (e.g., $+$, $-$), while non-terminals are variables that can be expanded into terminal or non-terminal symbols. For instance, the adopted BNF grammar for the **rastrigin** function approximation task is:

$\langle \text{expr} \rangle ::= \langle \text{expr} \rangle \langle \text{op} \rangle \langle \text{expr2} \rangle \mid \langle \text{expr2} \rangle$
$\langle \text{op} \rangle ::= + \mid - \mid \times$
$\langle \text{expr2} \rangle ::= x_1 \mid x_2 \mid \langle \text{value} \rangle$
$\langle \text{value} \rangle ::= \langle \text{digits} \rangle . \langle \text{digits} \rangle$
$\langle \text{digits} \rangle ::= \langle \text{digits} \rangle \langle \text{digit} \rangle \mid \langle \text{digit} \rangle$
$\langle \text{digit} \rangle ::= 0 \mid 1 \mid 2 \mid 3 \mid 4 \mid 5 \mid 6 \mid 7 \mid 8 \mid 9$

Each line of the grammar corresponds to a production rule, defining the symbols that replace a non-terminal. The angle brackets notation (< and >) is used to represent non-terminals. The "::=" symbol defines that the symbol on the left is to be replaced by the expression on the right. As for the "|" symbol, it separates the different replacement options, known as productions. Some non-terminals can have a recursive definition. For instance, a numeric value is defined by using recursive definition of $< digits >$, thus it can include a variable size of digits (the demonstration code exemplifies how the value 5.32 can be built). Examples of non-terminals include $< expr >$ and $< op >$, while x_1 and 0 are terminals. Similarly to the genetic programming demonstration, the **rastrigin** approximation grammar allows to define a wide range of arithmetic expressions, such as $3.6x_2 - 738.1$ or $2x_1x_2 + 66.37$.

To create the phenotype, the mapping procedure starts at the first symbol and ends when the program only includes terminals. For a currently analyzed gene integer (N) and production rule (with 0,1,...,R-1 replacement productions), the selected K-th production corresponds to the remainder of the division of N by R ($K = N \bmod R$). For the **rastrigin** approximation grammar, the genome $(2, 1, 0, 0, 4)$ produces the expression x_1+x_2 through the expansion: $< expr >$, $< expr >< op >< expr2 >$ (2mod2=0, first production of $< expr >$), $< expr2 >< op >< expr2 >$ (1mod2=1, second production of $< expr >$), $x_1 < op >< expr2 >$ (0mod3=0, first production of $< expr2 >$), $x_1 + < expr2 >$ (0mod3=0, first production of $< op >$), and $x_1 + x_2$ (4mod3=1, second production of $< expr2 >$). The demonstration code shows how this expression is built and also presents the expansion for the genome $(3, 2, 1, 1, 5, 0, 1, 3, 2)$.

Grammatical evolution can be implemented in R by using the **`gramEvol`** package (Noorian et al., 2016). The main **`gramEvol`** functions are:

- **`CreateGrammar`**—sets a BNF grammar using a list with **`gsrule`** (with strings) and **`grules`** (with R expressions);
- **`GrammarMap`**—expands a genome (sequence of integers) into an expression; and
- **`GrammaticalEvolution`**—runs a grammatical evolution using a grammar, evaluation function, and other parameters.

These functions are used in the **rastrigin** approximation task:

```
### rastrigin-ge.R ###

library(gramEvol) # load gramEvol

# auxiliary functions:
rastrigin=function(x) 10*length(x)+sum(x^2-10*cos(2*pi*x))
mse=function(y1,y2) mean((y1-y2)^2) # mean squared error

# set the grammar rules:
#   the first production rule is:
#   <expr> ::= <expr><op><expr2> | <expr2>
ruleDef=list(expr=gsrule("<expr><op><expr2>","<expr2>"),
             op=gsrule("+","-","*"),
             expr2=gsrule("x[1]","x[2]","<value>"),
             value=gsrule("<digits>.<digits>"),
             digits=gsrule("<digits><digit>","<digit>"),
             digit=grule(0,1,2,3,4,5,6,7,8,9)
            )
# create the BNF grammar object:
gDef=CreateGrammar(ruleDef)
```

```
# two expansion examples:
expr=GrammarMap(c(2,1,0,0,4),gDef,verbose=TRUE)
print(expr) # show expression: x[1] + x[2]
expr2=GrammarMap(c(3,2,1,1,5,0,1,3,2),gDef,verbose=TRUE)
print(expr2) # show expression: 5.32

# grammatical evolution setup:
Pop=100 # population size
Gen=100 # number of generations
Eli=1 # elitism
# simple monitoring function:
monitor=function(results)
{ # print(str(results)) shows all results components
  iter=results$population$currentIteration # current iteration
  f=results$best$cost # best fitness value
  if(iter==1||iter%%10==0) # show 1st and every 10 iter
    cat("iter:",iter,"f:",f,"\n")
}

# set the input samples (grid^2 data points):
grid=10 # size of the grid used
domain=matrix(ncol=2,nrow=grid^2) # 2D domain grid
domain[,1]=rep(seq(-5.2,5.2,length.out=grid),each=grid)
domain[,2]=rep(seq(-5.2,5.2,length.out=grid),times=grid)

y=apply(domain,1,rastrigin) # compute target output

eval1=function(x,expr) # x is an input vector with D=2
{ # expr can include x[1] or x[2] symbols
  eval(expr)
}

msevalue2=function(expr) # evaluation function
{
 y_expr=apply(domain,1,eval1,expr) # expr outputs for domain
 return (mse(y,y_expr))
}

set.seed(12345) # set for replicability
# run the grammar evolution:
ge=GrammaticalEvolution(gDef,msevalue2,optimizer="ga",
                        popSize=Pop,elitism=Eli,
                        iterations=Gen,monitorFunc=monitor)
b=ge$best # best solution
cat("evolved phenotype:")
print(b$expression)
cat("f:",b$cost,"\n")

# create approximation plot:
y2=apply(domain,1,eval1,b$expression)
MIN=min(y,y2);MAX=max(y,y2)
pdf("ge-function.pdf",width=7,height=7,paper="special")
plot(y,ylim=c(MIN,MAX),type="l",lwd=2,lty=1,
     xlab="points",ylab="function values")
```

```
lines(y2,type="l",lwd=2,lty=2)
legend("bottomright",leg=c("rastrigin","GE function"),lwd=2,
       lty=1:2)
dev.off()
```

The code contains several elements used by the genetic programming example. For instance, object **domain** includes the same two-dimensional input grid and the quality of the function approximation is measured by the same MSE function (**mse**, here explicitly defined since it is not included in **gramEvol**). An important difference is that a grammar needs to be defined, which is created using **CreateGrammar()**. This function requires a list (**ruleDef**) that contains the production rules. The previously presented BNF grammar is easily defined by using the **gsrule** and **grule** functions. The execution of the **GrammarMap** function is included to exemplify the expansion of the genomes $(2, 1, 0, 0, 4)$ and $(3, 2, 1, 1, 5, 0, 1, 3, 2)$. Another relevant difference, when compared with the genetic programming code, is the evaluation function **msevalue2**, since in this case each evaluated solution is an expression (the phenotype) and not a function (as in the genetic programming case). To evaluate each genome expansion, the **eval1** auxiliary function was created, which receives two input arguments (**x** and **expr**) and calls the **eval** and **expression()** R base functions. It should be noted that in **gramEvol**, the object **expr** that is inputted into the evaluation function is an expression (e.g., the first **expr** value of the executed code is **expression(3.7 + x[1] - x[2])**). A R console session is used to exemplify the **eval**, **eval1**, and **expression** functions:

```
> eval(1+2) # computes 1+2
[1] 3
> # expression allows to link c(1,2) with x[1] and x[2]
> eval1(c(1,2),expression(x[1]+x[2]))
[1] 3
```

Regarding the grammatical evolution, the algorithm is run by calling the **GrammaticalEvolution** function, in this case with a genetic algorithm optimizer (**optimizer="ga"**), a population size of **Pop=100**, elitism of **Eli=1**, stopped after **Gen=100** generations, and monitored every 10 iterations (**monitorFunc=monitor**). The code ends by showing the best evolved expression and creating a PDF file with the **rastrigin** function approximation plot. The execution result is:

```
> source("rastrigin-ge.R")
 Step Codon Symbol  Rule                Result
 0                  starting:           <expr>
 1    2     <expr>  <expr><op><expr2>   <expr><op><expr2>
 2    1     <expr>  <expr2>             <expr2><op><expr2>
 3    0     <expr2> x[1]                x[1]<op><expr2>
 4    0     <op>    +                   x[1]+<expr2>
 5    4     <expr2> x[2]                x[1]+x[2]
Valid Expression Found
x[1] + x[2]
```

```
 Step Codon Symbol    Rule                   Result
 0                    starting:              <expr>
 1    3      <expr>   <expr2>                <expr2>
 2    2      <expr2>  <value>                <value>
 3    1      <value>  <digits>.<digits> <digits>.<digits>
 4    1      <digits> <digit>                <digit>.<digits>
 5    5      <digit>  5                      5.<digits>
 6    0      <digits> <digits><digit>        5.<digits><digit>
 7    1      <digits> <digit>                5.<digit><digit>
 8    3      <digit>  3                      5.3<digit>
 9    2      <digit>  2                      5.32
Valid Expression Found
5.32
iter: 1 f: 110.5354
iter: 10 f: 110.5354
iter: 20 f: 88.75188
iter: 30 f: 88.75188
iter: 40 f: 88.75188
iter: 50 f: 87.1611
iter: 60 f: 87.11203
iter: 70 f: 87.11203
iter: 80 f: 87.11203
iter: 90 f: 87.11203
iter: 100 f: 87.11203
evolved phenotype:expression(x[1] * x[1] + x[2] - x[2] + 16 + x
    [2] * x[2])
f: 87.11203
```

The algorithm improved a solution from a MSE of 110.53 (generation 1) to 87.11 (generation 100). The best evolved expression is $x_1^2 + x_2^2 + 16$ (since $x_2 - x_2 = 0$) and the created PDF plot is shown in Fig. 5.14, denoting a close fit.

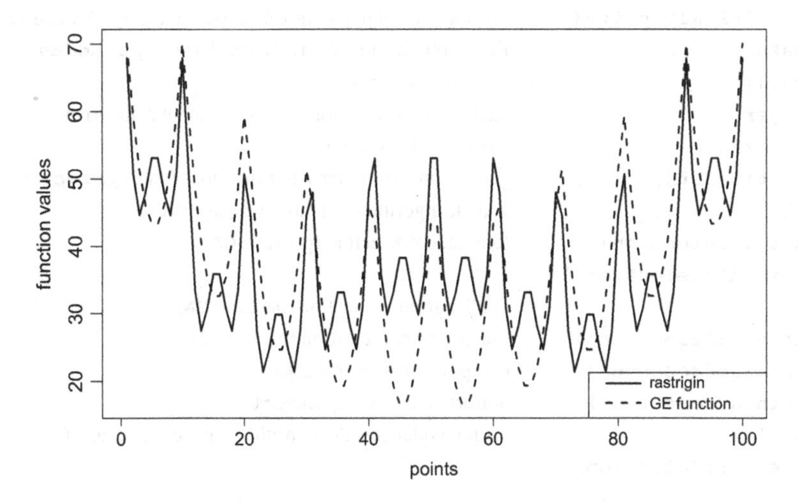

Fig. 5.14 Comparison of **rastrigin** function and best solution given by the grammatical evolution

5.12 Command Summary

CEDA()	implements EDAs based on multivariate copulas (package **copulaedas**)
CreateGrammar()	creates a grammar (package **gramEvol**)
DEoptim	package for differential evolution
DEopt()	differential evolution algorithm (package **NMOF**)
DEoptim()	differential evolution algorithm (package **DEoptim**)
DEoptim.control()	differential evolution control parameters (package **DEoptim**)
DEoptimR	package for differential evolution in pure R
GA	package with a general-purpose genetic algorithm
GAopt()	genetic algorithm function (package **NMOF**)
GrammarMap()	expands a sequence of integers (package **gramEvol**)
GrammaticalEvolution()	runs a grammatical evolution (package **gramEvol**)
JDEoptim()	differential evolution algorithm (package **DEoptimR**)
NMOF	package with numerical methods and optimization in finance
PSopt()	particle swarm optimization algorithm (package **NMOF**)
VEDA()	implements EDAs based on vines (package **copulaedas**)
abm.acor()	ant colony algorithm for continuous variables (package **evoper**)
constantFactorySet()	genetic programming set of constants (package **rgp**)
copulaedas	package for EDAs based on copulas
ecr	evolutionary computation in R package
ecr()	evolutionary computation function (package **ecr**)
edaCriticalPopSize()	sets the critical population size (package **copulaedas**)
edaRun()	EDA optimization algorithm (package **copulaedas**)
eval()	evaluates an R expression
evoper	package for estimation methods using metaheuristics
expression()	creates an R expression
functionSet()	genetic programming set of functions (package **rgp**)
ga()	genetic algorithm function (package **GA**)
ga_tourSelection()	tournament selection (package **GA**)
gabin_nlrSelection()	nonlinear rank selection (package **GA**)
gabin_rwSelection()	roulette wheel selection (package **GA**)
gabin_spCrossover()	one-point crossover (package **GA**)
gabin_uCrossover()	uniform crossover (package **GA**)
gaisl()	island model genetic algorithm function (package **GA**)
gaperm_lrSelection()	

	linear rank selection (package **GA**)
gareal_raMutation()	uniform random mutation (package **GA**)
genalg	package for genetic and evolutionary algorithms
geneticProgramming()	genetic programming algorithm (package **rgp**)
gramEvol	package for grammatical evolution
gray()	returns a vector of gray colors from a vector of gray levels
grule()	creates a production rule (package **gramEvol**)
gsrule()	creates a production rule (package **gramEvol**)
inputVariableSet()	genetic programming set of variables (package **rgp**)
match.call()	returns a call with all arguments
mcga	package with machine coded genetic algorithms
mcga()	machine coded genetic algorithm (package **mcga**)
mse()	mean squared error (package **rgp**)
mutBitflip()	bit mutation (package **ecr**)
mutGauss()	Gaussian mutation (package **ecr**)
mutPolynomial()	polynomial mutation (package **ecr**)
mutateSubtree()	random subtree mutation (package **rgp**)
plot.DEoptim()	plot differential evolution result (package **DEoptim**)
plot.rbga()	plot genetic/evolutionary algorithm result (package **genalg**)
points()	adds a sequence of points to a plot
pso	package for particle swarm optimization
psoptim	another package for particle swarm optimization
psoptim()	particle swarm optimization algorithm (package **pso**)
psoptim()	particle swarm optimization algorithm (package **psoptim**)
rbga()	evolutionary algorithm (package **genalg**)
rbga.bin()	genetic algorithm (package **genalg**)
recCrossover()	one-point crossover (package **ecr**)
recUnifCrossover()	uniform crossover (package **ecr**)
rgp	package for genetic programming
selGreedy()	greedy selection (package **ecr**)
selRoulette()	roulette wheel selection (package **ecr**)
selTournament()	tournament selection (package **ecr**)
setMethod()	creates and saves a method
setup()	wrapper of an operator (package **ecr**)
show()	shows an object
sinpi()	$sin(\pi \times x)$ trigonometric function
summary.DEoptim()	summarizes differential evolution result (package **DEoptim**)
summary.rbga()	summarizes genetic/evolutionary algorithm result (package **genalg**)
suppressWarnings()	evaluates its expression and ignores all warnings
try()	runs an expression that might fail and handles error-recovery
vignette()	views a particular vignette or lists available ones

5.13 Exercises

5.1 Apply the genetic algorithm implementations of packages **genalg**, **ecr**, **GA**, and **NMOF** to optimize the binary **sum of bits** task with $D = 20$. In all searches, use a population size of 20, an elitism of 1, a maximum of 100 iterations, and the other default values. Show the best optimized solutions and fitness values.

5.2 Consider the **eggholder** function ($D = 2$):

$$f = -(x_2 + 47)\sin\left(\sqrt{|x_2 + x_1/2 + 47|}\right) - x_1 \sin\left(\sqrt{|x_1 - x_2 + 47|}\right) \quad (5.6)$$

When the input range is set within $x_i \in [-512, 512]$, the optimum solution is $f(512, 404.2319) = -959.6407$. Adapt the code of file **compare2.R** (Sect. 5.7) such that three methods are compared to minimize the **eggholder** task: Monte Carlo search (Sect. 3.4), particle swarm optimization (SPSO 2011), and EDA (DVEDA). Use 10 runs for each method, with a maximum number of evaluations set to **MAXFN=1000** and solutions searched within the range [-512,512]. Consider the percentage of successes below -950. For the population based methods, use a population size of $N_P = 20$, and maximum number of iterations of $maxit = 50$.

5.3 Consider the original **bag prices** task ($D = 5$, Sect. 1.8) with a new hard constraint: $x_1 > x_2 > x_3 > x_4 > x_5$. Adapt the code of Sect. 5.8 in order to compare death-penalty and repair constraint handling strategies using an EDA of type UMDA. Hint: consider a simple repair solution that reorders each infeasible solution into a feasible one.

5.4 Consider the **eggholder** function of Exercise 5.2. Using the packages **genalg**, **GA** and **mcga**, compare the execution times and optimized value for the three functions **rbga** (sequential), **gaisl** (island parallel model with detected cores), and **mcga** (sequential). Use a population of $N_P = 400$ individuals, $maxit = 500$ iterations, and the other default parameters. Note: **mcga** does not bound the optimized solutions (just the initial population), thus a constraint handling strategy (e.g., death-penalty) should be adopted when running this algorithm.

5.5 Approximate the **eggholder** function of Exercise 5.2 using the Mean Absolute error (MAE), which is computed as $MAE = \sum_{i=1}^{N} |y_i - \hat{y}_i|/N$. Set the domain input with 100 samples randomly generated within the range $[-512, 512]$ and stop the algorithm after 50 iterations. Adapt two algorithms:

1. A genetic programming method with a population size of 100 and other default parameters. The genetic programming building blocks should be defined as:
   - function symbols—use the same functions/operators that appear at the **eggholder** equation;
   - constants—use a random sampling over the **eggholder** constants {2,47}; and
   - variables—two inputs (x_1 and x_2).

2. A grammatical evolution with a population size of 100 and other default parameters. The grammar should use the same functions/operations that appear at the **eggholder** equation.

Chapter 6
Multi-Objective Optimization

6.1 Introduction

In previous chapters, only single objective tasks were addressed. However, multiple goals are common in real-world domains. For instance, a company typically desires to increase sales while reducing production costs. Within its marketing department, the goal might include maximizing target audiences while minimizing the marketing budget. Also, in its production department, the same company might want to maximize the manufactured items, in terms of both quality and production numbers, while minimizing production time, costs, and waste of material. Often, the various objectives can conflict, where gaining in one goal involves losing in another one. Thus, there is a need to set the right trade-offs.

To handle multi-objective tasks, there are three main approaches (Freitas, 2004): weighted-formula, lexicographic, and Pareto front, whose R implementation details are discussed in the next sections, after presenting the demonstrative tasks selected for this chapter.

6.2 Multi-Objective Demonstrative Problems

This section includes three examples of simple multi-objective tasks that were selected to demonstrate the methods presented in this chapter. Given an D-dimensional variable vector $\mathbf{x} = \{x_1, \ldots, x_D\}$ the goal is to optimize a set of m objective functions $\{f_1(x_1, \ldots, x_D), \ldots, f_m(x_1, \ldots, x_D)\}$. To simplify the example demonstrations, only two $m = 2$ objective functions are adopted for each task.

The binary multi-objective goal consists in maximizing both functions of the set $\{f_{\text{sum of bits}}(x_1, \ldots, x_D), f_{\text{max sin}}(x_1, \ldots, x_D)\}$, where $x_i \in \{0, 1\}$ and the functions are defined in Eqs. 1.1 and 1.2. As explained in Sect. 1.8 (see also Fig. 1.5),

P. Cortez, *Modern Optimization with R*, Use R!,
https://doi.org/10.1007/978-3-030-72819-9_6

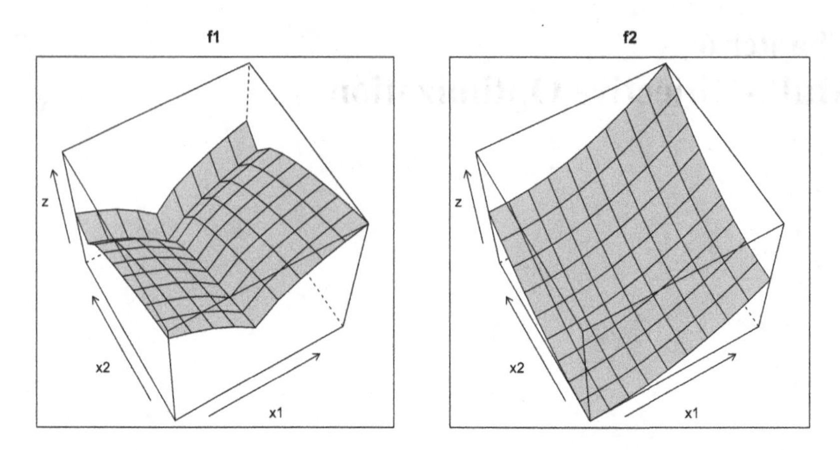

Fig. 6.1 Example of the FES1 f_1 (left) and f_2 (right) task landscapes ($D = 2$)

when $D = 8$ the optimum solutions are set at different point of the search space (**x**=(1,1,1,1,1,1,1,1) for f_1 and **x**=(1,0,0,0,0,0,0,0) for f_2), thus a trade-off is needed.

The **bag prices** integer multi-objective goal is set by maximizing f_1 and minimizing f_2, where $f_1 = f_{\text{bag prices}}$ and $f_2 = \sum_{i=1}^{D} sales(x_i)$, i.e., the number of bags that the factory will produce (see Sect. 1.8).

Finally, the real value multi-objective goal is defined in terms of the **FES1** benchmark (Huband et al., 2006), which involves minimizing both functions of the set:

$$\{f_1 = \sum_{i=1}^{D} |x_i - \exp((i/D)^2)/3|^{0.5},\ f_2 = \sum_{i=1}^{D} (x_i - 0.5\cos(10\pi i/D) - 0.5)^2\} \tag{6.1}$$

where $x_i \in [0, 1]$. As shown in Fig. 6.1, the minimum solutions are set at distinct points of the search space.

The R code related with the three multi-optimization tasks is presented in file **mo-tasks.R**:

```
### mo-tasks.R file ###

# binary multi-optimization goal:
sumbin=function(x) (sum(x))
intbin=function(x) sum(2^(which(rev(x==1))-1))
maxsin=function(x) # max sin (explained in Chapter 3)
{ D=length(x);x=intbin(x)
  return(sin(pi*(as.numeric(x))/(2^D))) }

# integer multi-optimization goal:
profit=function(x)       # x - a vector of prices
{ x=round(x,digits=0) # convert x into integer
  s=sales(x)             # get the expected sales
  c=cost(s)              # get the expected cost
```

```
  profit=sum(s*x-c)    # compute the profit
  return(profit)
}
cost=function(units,A=100,cpu=35-5*(1:length(units)))
{ return(A+cpu*units) }
sales=function(x,A=1000,B=200,C=141,
               m=seq(2,length.out=length(x),by=-0.25))
{ return(round(m*(A/log(x+B)-C),digits=0))}
produced=function(x) sum(sales(round(x)))

# real value FES1 benchmark:
fes1=function(x)
{ D=length(x);f1=0;f2=0
  for(i in 1:D)
    { f1=f1+abs(x[i]-exp((i/D)^2)/3)^0.5
      f2=f2+(x[i]-0.5*cos(10*pi/D)-0.5)^2
    }
  return(c(f1,f2))
}
```

6.3 Weighted-Formula Approach

The weighted-formula approach, also known as priori approach, has the advantage of being the simplest multi-objective solution, thus it is more easy to implement. This approach involves first assigning weights to each goal and then optimizing a quality Q measure that is typically set using an additive or multiplicative formula:

$$\begin{aligned} Q &= w_1 \times g_1 + w_2 \times g_2 + \ldots + w_n \times g_n \\ Q &= g_1^{w_1} \times g_1^{w_2} \times \ldots \times g_n^{w_n} \end{aligned} \tag{6.2}$$

where $g_1, g_2, \ldots, g_n$ denote the distinct goals and $w_1, w_2, \ldots, w_n$ the assigned weights.

As discussed in (Freitas, 2004; Konak et al., 2006), there several disadvantages with the weighted-formula approach. First, setting the ideal weights is often difficult and thus weights tend to be set ad hoc (e.g., based on intuition). Second, solving a problem to optimize Q for a particular vector $\mathbf{w}$ yields a single solution. This means that optimizing with a different combination of weights requires the execution of a new optimization procedure. Third, even if the weights are correctly defined, the search will miss trade-offs that might be interesting for the user. In particular, the linear combination of weights (as in the additive formula) limits the search for solutions in a non-convex region of the Pareto front (Fig. 6.2).

To solve the first two weighted-formula limitations, enhanced optimization variants have been proposed. One interesting example is the weight-based genetic algorithm (WBGA), which encodes a different weight vector into each solution of the genetic population (Konak et al., 2006).

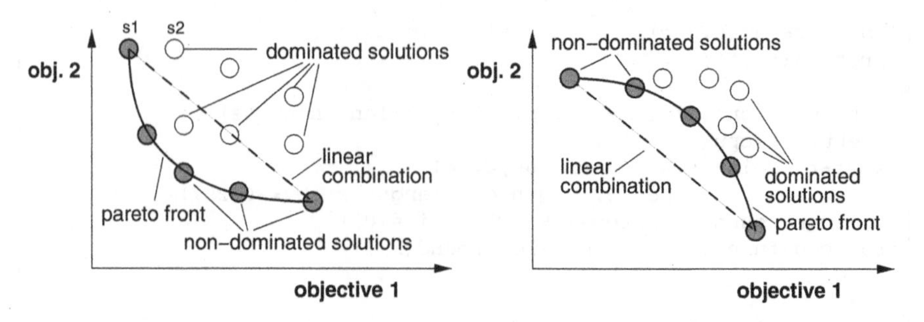

Fig. 6.2 Examples of convex (left) and non-convex (right) Pareto fronts, where the goal is to minimize both objectives 1 and 2

In this section, a pure weighted-formula approach is adopted for the three tasks presented in Sect. 6.2. Five additive weight combinations are tested: $\mathbf{w}_1 = (1.00, 0.00)$, $\mathbf{w}_2 = (0.75, 0.25)$, $\mathbf{w}_3 = (0.50, 0.50)$, $\mathbf{w}_4 = (0.75, 0.25)$ and $\mathbf{w}_5 = (0.00, 1.00)$. It should be noted that in all three tasks, there are different scales for each of the objectives (e.g., [0,8] range for $f_{\text{sum of bits}}$ and [0,1] for $f_{\text{max sin}}$). Thus, the optimization method will tend to improve more the objective associated with the largest scale, unless more differentiated weights are used. Nevertheless, for the sake of simplicity, the same weight combinations are used for all three benchmarks.

As the search engine, genetic and evolutionary algorithms are adopted, as implemented in the **genalg** package. This package can handle both binary (**rbga.bin** function) and real value (**rbga** function) representations. A distinct run of the optimization algorithm is executed for each of the five weight combinations. The population size is set to 20 individuals for **bag prices** and **FES1** multi-objective tasks, while a smaller population size of 12 is used for the simpler binary multi-objective problem. The weighted-formula R code is presented in file **wf-test.R**:

```
### wf-test.R file ###

source("MO-tasks.R") # load multi-optimization tasks
library(genalg) # load genalg package

set.seed(12345) # set for replicability

step=5 # number of weight combinations
w=matrix(ncol=2,nrow=step) # weight combinations
w[,1]=seq(1,0,length.out=step)
w[,2]=1-w[,1] # complementary weights (sum(w[i,])==1)

print("Weight combinations:")
print(w)

# --- binary task:
D=8 # 8 bits
# weighted evaluation function: W is a global vector
```

```
weval=function(x) return(W[1]*sumbin(x)+W[2]*maxsin(x))

cat("binary task:\n")
for(i in 1:step)
{
 W= -w[i,] # rbga.bin minimization goal: max. f1 and max. f2
 G=rbga.bin(size=D,popSize=12,iters=100,zeroToOneRatio=1,
            evalFunc=weval,elitism=1)
 b=G$population[which.min(G$evaluations),] # best individual
 cat("w",i,"best:",b)
 cat(" f=(",sumbin(b),",",round(maxsin(b),2),")","\n",sep="")
}

# --- integer task:
D=5 # 5 bag prices
# weighted evaluation function: W is a global vector
weval=function(x) return(W[1]*profit(x)+W[2]*produced(x))

cat("integer task:\n")
res=matrix(nrow=nrow(w),ncol=ncol(w)) # for CSV files
for(i in 1:step)
{
 W=c(-w[i,1],w[i,2]) # rbga min. goal: max. f1 and min. f2
 G=rbga(evalFunc=weval,stringMin=rep(1,D),stringMax=rep(1000,D),
       popSize=20,iters=100)
 b=round(G$population[which.min(G$evaluations),]) # best
 cat("w",i,"best:",b)
 cat(" f=(",profit(b),",",produced(b),")","\n",sep="")
 res[i,]=c(profit(b),produced(b))
}
write.table(res,"wf-bag.csv",
            row.names=FALSE,col.names=FALSE,sep=" ")

# --- real value task:
D=8 # dimension
# weighted evaluation function: W is a global vector
weval=function(x) return(sum(W*fes1(x)))

cat("real value task:\n")
for(i in 1:step)
{
 W=w[i,] # rbga minimization goal
 G=rbga(evalFunc=weval,stringMin=rep(0,D),stringMax=rep(1,D),
       popSize=20,iters=100)
 b=G$population[which.min(G$evaluations),] # best solution
 cat("w",i,"best:",round(b,2))
 cat(" f=(",round(fes1(b)[1],2),",",round(fes1(b)[2],2),")","\n",
     sep="")
 res[i,]=fes1(b)
}
write.table(res,"wf-fes1.csv",
            row.names=FALSE,col.names=FALSE,sep=" ")
```

The distinct weight combinations are stored in matrix **w**. Given that the **genalg** package performs a minimization, the $f'(s) = -f(s)$ transformation (Sect. 1.4) is adopted when the objective requires a maximization and thus the auxiliary **W** vector is used to multiple the weight values by -1 when needed. After executing each optimization run, the code displays the best evolved solution and also the two objective evaluation values. For comparison with other multi-objective approaches, the best evaluation values are stored into CSV files (using the **write.table** function) for the last two tasks. The execution result is:

```
> source("wf-test.R")
[1] "Weight combinations:"
     [,1] [,2]
[1,] 1.00 0.00
[2,] 0.75 0.25
[3,] 0.50 0.50
[4,] 0.25 0.75
[5,] 0.00 1.00
binary task:
w 1 best: 1 1 1 1 1 1 1 1 f=(8,0.01)
w 2 best: 1 1 1 1 1 1 1 1 f=(8,0.01)
w 3 best: 1 1 1 1 1 1 1 1 f=(8,0.01)
w 4 best: 0 1 1 1 1 1 1 1 f=(7,1)
w 5 best: 0 1 1 1 1 1 1 1 f=(7,1)
integer task:
w 1 best: 375 401 408 352 328 f=(43417,124)
w 2 best: 426 376 419 411 370 f=(43585,116)
w 3 best: 366 414 374 388 446 f=(43452,119)
w 4 best: 425 432 425 413 371 f=(43739,112)
w 5 best: 996 990 999 981 933 f=(423,1)
real value task:
w 1 best: 0.33 0.35 0.38 0.42 0.48 0.59 0.7 0.9 f=(0.67,1.4)
w 2 best: 0.34 0.36 0.4 0.43 0.51 0.59 0.72 0.9 f=(0.66,1.46)
w 3 best: 0.34 0.36 0.39 0.4 0.49 0.57 0.71 0.9 f=(0.57,1.39)
w 4 best: 0.34 0.34 0.37 0.43 0.49 0.31 0.26 0.25 f=(2.34,0.37)
w 5 best: 0.16 0.16 0.15 0.15 0.13 0.15 0.16 0.16 f=(4.75,0)
```

As expected, the obtained results show that in general the genetic and evolutionary algorithms manage to get the best f_1 values for the $\mathbf{w}_1 = (1.00, 0.00)$ weight combination and best f_2 values for the $\mathbf{w}_5 = (0.00, 1.00)$ vector of weights. The quality of the remaining task evolved solutions (for **bag prices** and **FES1**) will be discussed in Sect. 6.5.

6.4 Lexicographic Approach

Under the lexicographic approach, different priorities are assigned to different objectives, such that the objectives are optimized in their priority order (Freitas, 2004). When two solutions are compared, first the evaluation measure for the highest-priority objective is compared. If the first solution is significantly better (e.g., using a given tolerance value) than the second solution, then the former is chosen. Else, the comparison is set using the second highest-priority objective. The process is repeated until a clear winner is found. If there is no clear winner, then the solution with the best highest-priority objective can be selected.

In (Freitas, 2004), the advantages and disadvantages of the lexicographic approach are highlighted. When compared with the weighted-formula, the lexicographic approach has the advantage of avoiding the problem of mixing non-commensurable criteria in the same formula, as it treats each criterion separately. Also, if the intention is to just to compare several solutions, then the lexicographic approach is easier when compared with the Pareto approach. However, the lexicographic approach requires the user to a priori define the criteria priorities and tolerance thresholds, which similarly to the weighted-formula are set ad hoc.

Given that in previous section an evolutionary/genetic algorithm was adopted, the presented lexicographic implementation also adopts the same base algorithm. In particular, the **rbga.bin()** function code is adapted by replacing the probabilistic nonlinear rank selection with a tournament selection (Sect. 5.2). The advantage of using tournament is that there is no need for a single fitness value, since the selection of what is the "best" can be performed under a lexicographic comparison with the k solutions. It should be noted that the same tournament function could be used to get other multi-objective optimization adaptations. For instance, a lexicographic hill climbing could easily be achieved by setting the *best* function of Algorithm 2 as the same tournament operator (in this case by comparing $k = 2$ solutions).

The lexicographic genetic algorithm R code is provided in the file **lg-ga.R**:

```
### lg-ga.R file ###

# lexicographic comparison of several solutions:
#    x - is a matrix with several objectives at each column
#        and each row is related with a solution
lexibest=function(x) # assumes LEXI is defined (global variable)
{
 size=nrow(x); m=ncol(x)
 candidates=1:size
 stop=FALSE; i=1
 while(!stop)
   {
    F=x[candidates,i] # i-th goal
    minFID=which.min(F) # minimization goal is assumed
    minF=F[minFID]
    # compute tolerance value
    if(minF>-1 && minF<1) tolerance=LEXI[i]
    else tolerance=abs(LEXI[i]*minF)
    I=which((F-minF)<=tolerance)
    if(length(I)>0) # at least one candidate
      candidates=candidates[I] # update candidates
    else stop=TRUE
    if(!stop && i==m) stop=TRUE
    else i=i+1
   }
```

```
 if(length(candidates)>1)
   { # return highest priority goal if no clear winner:
     stop=FALSE; i=1
     while(!stop)
       {
        minF=min(x[candidates,i])
        I=which(x[candidates,i]==minF)
        candidates=candidates[I]
        if(length(candidates)==1||i==m) stop=TRUE
        else i=i+1
       }
     # remove (any) extra duplicate individuals:
     candidates=candidates[1]
   }
 # return lexibest:
 return(candidates)
}

# compare k randomly selected solutions from Population:
#    returns n best indexes of Population (decreasing order)
#    m is the number of objectives
tournament=function(Population,evalFunc,k,n,m=2)
{
 popSize=nrow(Population)
 PID=sample(1:popSize,k) # select k random tournament solutions
 E=matrix(nrow=k,ncol=m) # evaluations of tournament solutions
 for(i in 1:k) # evaluate tournament
    E[i,]=evalFunc(Population[PID[i],])

 # return best n individuals:
 B=lexibest(E); i=1; res=PID[B] # best individual
 while(i<n) # other best individuals
   {
     E=E[-B,];PID=PID[-B] # all except B
     if(is.matrix(E)) B=lexibest(E)
     else B=1 # only 1 row
     res=c(res,PID[B])
     i=i+1
   }
 return(res)
}

# lexicographic adapted version of rbga.bin:
#   this function is almost identical to rbga.bin except that
#   the code was simplified and a lexicographic tournament is
#   used instead of roulette wheel selection
lrbga.bin=function(size=10, suggestions=NULL, popSize=200,
                   iters=100, mutationChance=NA, elitism=NA,
                   zeroToOneRatio=10,evalFunc=NULL)
{
 vars=size
 if(is.na(mutationChance)) { mutationChance=1/(vars + 1) }
 if(is.na(elitism)) { elitism=floor(popSize/5)}
 if(!is.null(suggestions))
```

```
  {
   population=matrix(nrow=popSize, ncol=vars)
   suggestionCount=dim(suggestions)[1]
   for(i in 1:suggestionCount)
     population[i, ]=suggestions[i, ]
   for(child in (suggestionCount + 1):popSize)
     {
      population[child, ]=sample(c(rep(0, zeroToOneRatio),1),
          vars,rep=TRUE)
      while(sum(population[child, ])==0)
         population[child, ]=sample(c(rep(0, zeroToOneRatio),1),
             vars,rep=TRUE)
     }
  }
else
  {
   population=matrix(nrow=popSize, ncol=vars)
   for(child in 1:popSize)
     {
      population[child,]=sample(c(rep(0, zeroToOneRatio),1),vars
          ,rep=TRUE)
      while (sum(population[child, ]) == 0)
        population[child, ]=sample(c(rep(0, zeroToOneRatio),1),
            vars,rep=TRUE)
     }
  }
# main GA cycle:
for(iter in 1:iters)
  {
   newPopulation=matrix(nrow=popSize, ncol=vars)
   if(elitism>0) # applying elitism:
     {
      elitismID=tournament(population,evalFunc,k=popSize,n=
          elitism)
      newPopulation[1:elitism,]=population[elitismID,]
     }
   #  applying crossover:
   for(child in (elitism + 1):popSize)
     {
      ### very new code inserted here : ###
      pID1=tournament(population,evalFunc=evalFunc,k=2,n=1)
      pID2=tournament(population,evalFunc=evalFunc,k=2,n=1)
      parents=population[c(pID1,pID2),]
      ### end of very new code          ###
      crossOverPoint=sample(0:vars, 1)
      if(crossOverPoint == 0)
        newPopulation[child,]=parents[2,]
      else if(crossOverPoint == vars)
        newPopulation[child, ]=parents[1, ]
      else
        {
         newPopulation[child,]=c(parents[1,][1:crossOverPoint],
             parents[2,][(crossOverPoint+1):vars])
         while(sum(newPopulation[child,])==0)
```

```
            newPopulation[child, ]=sample(c(rep(0,zeroToOneRatio)
                ,1),vars,rep=TRUE)
          }
        }
      population=newPopulation # store new population
      if(mutationChance>0) # applying mutations:
        {
         mutationCount=0
         for(object in (elitism+1):popSize)
           {
            for(var in 1:vars)
              {
               if(runif(1)< mutationChance)
                 {
                  population[object, var]=sample(c(rep(0,
                      zeroToOneRatio),1),1)
                  mutationCount=mutationCount+1
                 }
              }
           }
        }
    } # end of GA main cycle
 result=list(type="binary chromosome",size=size,popSize=popSize,
       iters=iters,suggestions=suggestions,
       population=population,elitism=elitism,
       mutationChance=mutationChance)
 return(result)
}
```

The new **lrbga.bin()** is a simplified version of the **rbga.bin()** function (which is accessible by simple typing **> rbga.bin** in the R console), where all verbose and monitoring code have been removed. The most important change is that before applying the crossover a tournament with k=2 individuals is set to select the two parents. It should be noted that $k = 2$ is the most popular tournament strategy (Michalewicz and Fogel, 2004). The same tournament operator is also used to select the elitism individuals from the population. In this case, $k = N_p$ (the population size) and $n = E$ (elitism number). The **tournament()** function returns the n best individuals, according to a lexicographic comparison. Under this implementation, the evaluation function needs to return vector with the fitness values for all m objectives.

The **tournament()** assumes the first objective as the highest-priority function, the second objective is considered the second highest-priority function, and so on. It should be noted that **tournament()** uses the **is.matrix()** R function, which returns true if **x** is a matrix object. The lexicographic comparison is only executed when there are two or more solutions (which occurs when **x** object is a matrix). Function **lexibest()** implements the lexicographic comparison, returning the best index of the tournament population. This function assumes that the tolerance thresholds are defined in object **LEXI**. Also, these thresholds are interpreted as percentages if $-1 < f_i < 1$ for the i-th objective, else absolute values are used.

Working from the highest priority to the smallest one, the tournament population is reduced on a step by step basis, such that on each iteration only the best solutions within the tolerance range for the i-th objective are selected. If there is no clear winner, **lexibest()** selects the best solution, as evaluated from the highest to the smallest priority objective.

The optimization of the binary multi-objective goal is coded in file **lg-test.R**, using a tolerance of 20% for both objectives and the same other parameters that were used in Sect. 6.3:

```
### lg-test.R file ###

source("mo-tasks.R") # load multi-optimization tasks
source("lg-ga.R") # load lrgba.bin
set.seed(12345) # set for replicability

LEXI=c(0.2,0.2) # tolerance 20% for each goal
cat("tolerance thresholds:",LEXI,"\n")

# --- binary task:
D=8 # 8 bits
# mineval: transform binary objectives into minimization goal
#         returns a vector with 2 values, one per objective:
mineval=function(x) return(c(-sumbin(x),-maxsin(x)))
popSize=12
G=lrbga.bin(size=D,popSize=popSize,iters=100,zeroToOneRatio=1,
            evalFunc=mineval,elitism=1)
print("Ranking of last population:")
B=tournament(G$population,mineval,k=popSize,n=popSize,m=2)
for(i in 1:popSize)
{
 x=G$population[B[i],]
 cat(x," f=(",sumbin(x),",",round(maxsin(x),2),")","\n",sep="")
}
```

Given that there is not a single best solution, after executing the lexicographic genetic algorithm, the code shows a ranking (according to the lexicographic criterion) of all individuals from last population:

```
> source("lg-test.R")
tolerance thresholds: 0.2 0.2
[1] "Ranking of last population:"
01111111 f=(7,1)
01111111 f=(7,1)
01111111 f=(7,1)
01111111 f=(7,1)
01111111 f=(7,1)
01111111 f=(7,1)
01111110 f=(6,1)
01111101 f=(6,1)
01110111 f=(6,0.99)
01110111 f=(6,0.99)
01011111 f=(6,0.92)
00111111 f=(6,0.7)
```

With a single run, the lexicographic algorithm is capable of finding the same $(f_1, f_2) = (7, 1)$ solution that belongs to the Pareto front (see next section).

6.5 Pareto Approach

A solution s_1 dominates (in the Pareto sense) a solution s_2 if s_1 is better than s_2 in one objective and as least as good as s_2 in all other objectives. A solution s_i is non-dominated when there is no solution s_j that dominates s_i and the Pareto front contains all non-dominated solutions (Luke, 2015). The left of Fig. 6.2 exemplifies these concepts, where s_1 is a non-dominated solution and part of the Pareto front, while s_2 is a dominated one (both solutions have the same f_2 value but s_1 presents a better f_1). Assuming this concept, Pareto multi-objective optimization methods return a set of non-dominated solutions (from the Pareto front), rather than just a single solution.

When compared with previous approaches (weighted-formula and lexicographic), the Pareto multi-objective optimization presents several advantages (Freitas, 2004). It is a more natural method, since a "true" multi-objective approach is executed, providing to the user an interesting set of distinct solutions and letting the user (a posteriori) to decide which one is best. Moreover, under a single execution, the method optimizes the distinct objectives, thus no multiple runs are required to get the Pareto front points. In addition, there is no need to set ad hoc weights or tolerance values. The drawback of the Pareto approach is that a larger search space needs to be explored and tracked, thus Pareto based methods tend to be more complex than single objective counterparts.

Considering that Pareto based methods need to keep track of a population of solutions, evolutionary algorithms have become a natural and popular solution to generate Pareto optimal solutions. Thus this type of search, often termed Multi-Objective Evolutionary Algorithms (MOEA) (Xu et al., 2020), Evolutionary Multi-objective Optimization Algorithms (EMOA) (Beume et al., 2007), or Evolutionary Multi-criteria Optimization (EMO), is detailed in this section. Nevertheless, it should be noted that the Pareto approach can also be adapted to other population based metaheuristics, such as the particle swarm optimization algorithm (Trivedi et al., 2020). The adaptations often involve storing a Pareto list of best solutions and employing heuristics to preserve diversity of the solutions.

Due to the importance of multi-objective algorithms, several MOEAs have been proposed, such as: Strength Pareto Evolutionary Algorithm 2 (SPEA-2) (Deb, 2001); Non-dominated Sorting Genetic Algorithm, versions II (NSGA-II) (Deb, 2001) and III (NSGA-III) (Deb and Jain, 2014); S Metric Selection EMOA (SMS-EMOA) (Beume et al., 2007); and Aspiration Set EMOA (AS-EMOA) (Rudolph et al., 2014). MOEA often use Pareto based ranking schemes, where individuals in the Pareto front are rank 1, then the front solutions are removed and the individuals from the new front are rank 2, and so on.

Different Pareto based optimization methods can be compared visually, particularly if the number of distinct objectives is relatively small (e.g., $m < 4$, allowing an easier two- or three-dimensional analysis). Nevertheless, there are also measures that compute a single value, such as the hypervolume indicator (also known as S metric or Lebesgue measure) (Beume et al., 2009). This indicator requires the definition of a reference point, which often is set as the worst solution (an example is plotted in Fig. 6.4). The indicator computes the hyper dimensional volume (or area if $m = 2$) of the Pareto front that is set within a hyper cube that is defined by the reference point. The hypervolume can also be used to guide the acceptance of the searched solutions, such as adopted by the SMS-EMOA algorithm (Beume et al., 2007).

This section first focuses on the **mco** package, which is one of the first MOEA R packages (it dates back to 2008). The package implements the popular NSGA-II algorithm, an evolutionary algorithm variant specifically designed for multi-objective optimization and that uses three useful concepts: Pareto front ranking, elitism and sparsity. The full NSGA-II algorithmic details can be found in (Deb, 2001) and (Luke, 2015), although the skeleton of the algorithm is similar to Algorithm 5. The initial population P is randomly generated and then a cycle is executed until a termination criterion is met. Within each cycle, NSGA-II uses a Pareto ranking scheme to assign a ranking number to each individual of the population. An elitism scheme is also adopted, storing an archive (P_E) of the best individuals. The elitism individuals are selected taking into account their rank number and also their sparsity. An individual is in a sparse region if the neighbor individuals are not too close to it. To measure how close two points are, a distance metric is used, such as Manhattan distance, which is defined by the sum of all m objective differences between the two points. Then, a new population $Children \leftarrow breed(P_E)$ is created, often by using a tournament selection (e.g., with $k = 2$), where $breed$ is a function that employs crossover and mutation operators. The next population is set as the union of the new solutions and the archive ($P \leftarrow Children \cup P_E$) and a new cycle is executed over P.

In the **mco** package, NSGA-II is implemented with the **nsga2** function. The package also contains other functions, such as related with multi-objective benchmarks (e.g., **belegundu()**) and Pareto front (e.g., **paretoSet()**). The useful **nsga2()** function performs a minimization of vectors of real numbers and includes the main parameters:

- **fn**—function to be minimized (should return a vector with the several objective values);
- **idim**—input dimension (D);
- **odim**—output dimension (number of objective functions, m);
- **...**—extra arguments to be passed **fn**;
- **lower.bounds**, **upper.bounds**—lower and upper bounds;
- **popsize**—population size (N_P, default is 100);
- **generations**—number of generations ($maxit$, default is 100) or a vector;
- **cprob**—crossover probability (default is 0.7); and
- **mprob**—mutation probability (default is 0.2).

The function returns a list with the final population (if **generations** is a number), with components:

- **$par**—the population values;
- **$value**—matrix with the best objective values (in columns) for the last population individuals (in rows); and
- **$pareto.optimal**—a Boolean vector that indicates which individuals from the last generation belong to the Pareto front.

When **generations** is a vector, a vector list is returned where the i-th element contains the population after **generations[i]** iterations (an R code example is shown in the next presented code).

File **ngsa-test.R** codes the NSGA-II optimization of the three multi-objective tutorial tasks:

```
### nsga2-test.R file ###

source("mo-tasks.R") # load multi-optimization tasks
library(mco) # load mco package

set.seed(12345) # set for replicability
m=2 # two objectives

# --- binary task:
D=8 # 8 bits
# eval: transform binary objectives into minimization goal
#       round(x) is used to convert real number to 0 or 1 values
beval=function(x) c(-sumbin(round(x)),-maxsin(round(x)))

cat("binary task:\n")
G=nsga2(fn=beval,idim=D,odim=m,
        lower.bounds=rep(0,D),upper.bounds=rep(1,D),
        popsize=12,generations=100)
# show last Pareto front
I=which(G$pareto.optimal)
for(i in I)
{
 x=round(G$par[i,])
 cat(x," f=(",sumbin(x),",",round(maxsin(x),2),")","\n",sep="")
}

# --- integer task:
D=5 # 5 bag prices
# ieval: integer evaluation (minimization goal):
ieval=function(x) c(-profit(x),produced(x))
# function that sorts matrix m according to 1st column
o1=function(m) # m is a matrix
  return(m[order(m[,1]),])

cat("integer task:\n")
G=nsga2(fn=ieval,idim=5,odim=m,
        lower.bounds=rep(1,D),upper.bounds=rep(1000,D),
        popsize=20,generations=1:100)
```

```
# show best individuals:
I=which(G[[100]]$pareto.optimal)
for(i in I)
{
 x=round(G[[100]]$par[i,])
 cat(x," f=(",profit(x),",",produced(x),")","\n",sep=" ")
}
# create PDF with Pareto front evolution:
pdf(file="nsga-bag.pdf",paper="special",height=5,width=5)
par(mar=c(4.0,4.0,0.1,0.1))
I=1:100
for(i in I)
{ P=G[[i]]$value # objectives f1 and f2
  P[,1]=-1*P[,1] # show positive f1 values
  # color from light gray (75) to dark (1):
  COL=paste("gray",round(76-i*0.75),sep="")
  if(i==1) plot(P,xlim=c(-500,44000),ylim=c(0,140),
                xlab="f1",ylab="f2",cex=0.5,col=COL)
  Pareto=P[G[[i]]$pareto.optimal,]
  # sort Pareto according to x axis:
  Pareto=o1(Pareto)
  points(P,type="p",pch=1,cex=0.5,col=COL)
  lines(Pareto,type="l",cex=0.5,col=COL)
}
dev.off()

# create PDF comparing NSGA-II with WF:
pdf(file="nsga-bag2.pdf",paper="special",height=5,width=5)
par(mar=c(4.0,4.0,0.1,0.1))
# NSGA-II best results:
P=G[[100]]$value # objectives f1 and f2
P[,1]=-1*P[,1] # show positive f1 values
Pareto=P[G[[100]]$pareto.optimal,]
# sort Pareto according to x axis:
Pareto=o1(Pareto)
plot(Pareto,xlim=c(-500,44000),ylim=c(0,140),
     xlab="f1",ylab="f2",type="b",lwd=2,lty=1,pch=1)
# weight-formula (wf) best results:
wf=read.table("wf-bag.csv",sep=" ") # data.frame
# paretoFilter only works with minimization goals:
wf=as.matrix(cbind(-wf[,1],wf[,2])) # matrix with -f1,f2
pwf=paretoFilter(wf) # get the Pareto front points of wf
wf[,1]=-wf[,1] # set to the f1,f2 domain
pwf[,1]=-pwf[,1] # set to the f1,f2 domain
points(wf,pch=3,lwd=2) # plot all wf points
lines(pwf,type="l",lty=2,lwd=2)
legend("topleft",c("NSGA-II","weighted-formula"),
       lwd=2,lty=1:2,pch=c(1,3))
dev.off()

# --- real value task:
D=8 # dimension
cat("real value task:\n")
G=nsga2(fn=fes1,idim=D,odim=m,
```

```
         lower.bounds=rep(0,D),upper.bounds=rep(1,D),
         popsize=20,generations=1:100)
# show best individuals:
I=which(G[[100]]$pareto.optimal)
for(i in I)
{
 x=round(G[[100]]$par[i,],digits=2); cat(x)
 cat(" f=(",round(fes1(x)[1],2),",",round(fes1(x)[2],2),")","\n",
     sep="")
}
# create PDF with Pareto front evolution:
pdf(file="nsga-fes1.pdf",paper="special",height=5,width=5)
par(mar=c(4.0,4.0,0.1,0.1))
I=1:100
for(i in I)
{ P=G[[i]]$value # objectives f1 and f2
  # color from light gray (75) to dark (1):
  COL=paste("gray",round(76-i*0.75),sep="")
  if(i==1) plot(P,xlim=c(0.5,5.0),ylim=c(0,2.0),
                xlab="f1",ylab="f2",cex=0.5,col=COL)
  Pareto=P[G[[i]]$pareto.optimal,]
  # sort Pareto according to x axis:
  Pareto=o1(Pareto)
  points(Pareto,type="p",pch=1,cex=0.5,col=COL)
  lines(Pareto,type="l",cex=0.5,col=COL)
}
dev.off()

# create PDF comparing NSGA-II with WF:
pdf(file="nsga-fes1-2.pdf",paper="special",height=5,width=5)
par(mar=c(4.0,4.0,0.1,0.1))
# NSGA-II best results:
P=G[[100]]$value # objectives f1 and f2
Pareto=P[G[[100]]$pareto.optimal,]
# sort Pareto according to x axis:
Pareto=o1(Pareto)
plot(Pareto,xlim=c(0.5,5.0),ylim=c(0,2.0),
     xlab="f1",ylab="f2",type="b",lwd=2,pch=1)
# weight-formula best results:
wf=read.table("wf-fes1.csv",sep=" ") # data.frame
wf=as.matrix(wf) # convert to matrix
pwf=paretoFilter(wf) # get the Pareto front points of wf
points(wf,pch=3,lwd=2) # plot all wf points
lines(pwf,type="l",lty=2,lwd=2)
legend("top",c("NSGA-II","weighted-formula"),
       lwd=2,lty=1:2,pch=c(1,3))
dev.off()
```

The execution of function **nsga2** is straightforward taking into account the weight-formula and lexicographic examples. For the binary task, each solution parameter ($\in [0, 1]$) is first rounded in order to transform it into a binary number, since **nsga2()** only works with real values. After calling the algorithm, the code shows all Pareto front solutions from the last generation. For each of the **bag prices** and

FES1 tasks, the code also creates two PDF files. The first PDF contains the search evolution in terms of the f_1 (x-axis) and f_2 (y-axis) objectives, where individual solutions are represented by small circle points and the Pareto front solutions are connected with lines. Also, a varying color scheme is adopted to plot the points and lines, ranging from light gray (first generation) to black (last generation). Before plotting each Pareto front (via the **points** and **lines** functions), the matrix object **Pareto** is first sorted according to the first objective. This is achieved by using the auxiliary **o1** function, which uses the R base function **order()** that rearranges a matrix. The second PDF compares the best Pareto front optimized by NSGA-II with the five solutions obtained by the five runs (with different weight combinations) executed for the weighted-formula approach. Since some of the five solutions can be dominated, the **paretoFilter** function of the **mco** package is used to discard dominated solutions from the weighted-formula search curve (shown using a dashed line). The **paretoFilter** function only works with minimization goals (similarly to the **nsga2** usage). Thus, the f_1 integer bag price results are first transformed into their negative form ($-f_1$) before calling the **paretoFilter** code. The execution result is:

```
> source("nsga2-test.R")
binary task:
01111111 f=(7,1)
11111111 f=(8,0.01)
01111111 f=(7,1)
11111111 f=(8,0.01)
01111111 f=(7,1)
11111111 f=(8,0.01)
11111111 f=(8,0.01)
01111111 f=(7,1)
11111111 f=(8,0.01)
01111111 f=(7,1)
01111111 f=(7,1)
11111111 f=(8,0.01)
integer task:
396 457 391 413 367  f=( 43317 , 114 )
1000 991 991 981 975  f=( -500 , 0 )
993 960 991 704 895  f=( 7028 , 10 )
995 999 999 850 965  f=( 2960 , 4 )
620 465 380 439 664  f=( 39046 , 86 )
818 790 847 778 995  f=( 17494 , 23 )
988 674 777 755 708  f=( 20458 , 30 )
608 704 797 763 963  f=( 26219 , 40 )
870 890 828 773 975  f=( 14125 , 18 )
620 361 380 439 664  f=( 39446 , 94 )
510 573 607 598 589  f=( 38563 , 72 )
396 457 426 413 367  f=( 42939 , 111 )
594 663 710 778 876  f=( 29065 , 46 )
823 573 607 590 589  f=( 32986 , 56 )
618 590 607 641 589  f=( 36424 , 63 )
793 590 679 764 582  f=( 30876 , 49 )
616 590 679 764 589  f=( 34211 , 57 )
466 561 449 413 367  f=( 42159 , 99 )
```

```
944 984 963 772 769  f=( 10051 , 13 )
818 649 724 778 982  f=( 22575 , 33 )
real value task:
0.34 0.35 0.39 0.44 0.49 0.59 0.7 0.9 f=(0.63,1.41)
0.15 0.15 0.17 0.14 0.12 0.16 0.11 0.16 f=(4.79,0)
0.34 0.32 0.38 0.44 0.49 0.48 0.18 0.3 f=(2.28,0.46)
0.34 0.35 0.38 0.44 0.49 0.58 0.18 0.3 f=(1.91,0.55)
0.34 0.35 0.38 0.44 0.5 0.59 0.58 0.9 f=(0.88,1.3)
0.34 0.35 0.34 0.44 0.49 0.58 0.71 0.75 f=(1.03,1.19)
0.34 0.35 0.38 0.44 0.49 0.58 0.7 0.62 f=(1.06,1.06)
0.19 0.28 0.21 0.14 0.12 0.22 0.14 0.14 f=(4.46,0.03)
0.34 0.35 0.39 0.4 0.37 0.17 0.17 0.17 f=(2.95,0.25)
0.34 0.35 0.39 0.44 0.49 0.58 0.21 0.51 f=(1.76,0.67)
0.17 0.35 0.31 0.26 0.38 0.26 0.16 0.2 f=(3.65,0.15)
0.34 0.23 0.17 0.31 0.11 0.16 0.1 0.18 f=(4.1,0.08)
0.34 0.35 0.38 0.44 0.49 0.58 0.7 0.5 f=(1.17,0.96)
0.22 0.13 0.38 0.22 0.2 0.28 0.22 0.24 f=(3.95,0.1)
0.33 0.29 0.38 0.44 0.49 0.39 0.18 0.27 f=(2.54,0.39)
0.21 0.35 0.38 0.26 0.37 0.26 0.22 0.19 f=(3.37,0.18)
0.34 0.36 0.34 0.44 0.49 0.58 0.64 0.49 f=(1.47,0.87)
0.34 0.36 0.38 0.43 0.5 0.58 0.7 0.38 f=(1.23,0.89)
0.34 0.35 0.39 0.44 0.49 0.58 0.69 0.91 f=(0.64,1.41)
0.36 0.37 0.38 0.27 0.49 0.26 0.18 0.27 f=(2.88,0.31)
```

For the sake of simplicity, the comparison of NSGA-II results with other approaches is performed here using only a single NGSA-II run. For a more robust comparison, several runs should be applied, as shown by the last **fes1** demonstration presented in this section.

The binary multi-objective task is quite simple given that the optimum Pareto front only contains two solutions (found by NSGA-II): $(f_1, f_2) = (7, 1)$ and $(f_1, f_2) = (8, 0.01)$. When compared with the weight-formula, the same best solutions were obtained, although NSGA-II gets both solutions under the same run.

As shown in the left two graphs of Fig. 6.3, the **bag prices** and **FES1** are more complex multi-objective tasks when compared with the binary **sum of bits/max sin** problem. Figure 6.3 reveals that the optimum Pareto fronts seem to have a convex shape for both integer and real value optimization tasks, although the final shape of such Pareto curve is only achieved during the last generations of NSGA-II. The right plots of Fig. 6.3 confirm that NSGA-II optimizes within a single run a more interesting Pareto front when compared with the results obtained by the weighted-formula approach (and that require executing 5 runs). For **bag prices**, the weighted approach misses several interesting solutions outside the linear weight combination (dashed line) trade-offs.

The **bag prices** and **fes1** multi-objective tasks are further selected to demonstrate other MOEA R packages: **`ecr`** (presented in Sect. 5.2) and **`MaOEA`** (Irawan and Naujoks, 2019). These packages implement additional MOEA algorithms, namely: SMS-EMOA and AS-EMOA—package **`ecr`**; and NSGA-III—package **`MaOEA`**. The SMS-EMOA works by selecting in each generation two parents uniformly, applying crossover and mutation operators and selecting the best solutions that maximize the hypervolume measure (Beume et al., 2007). To compute the indicator,

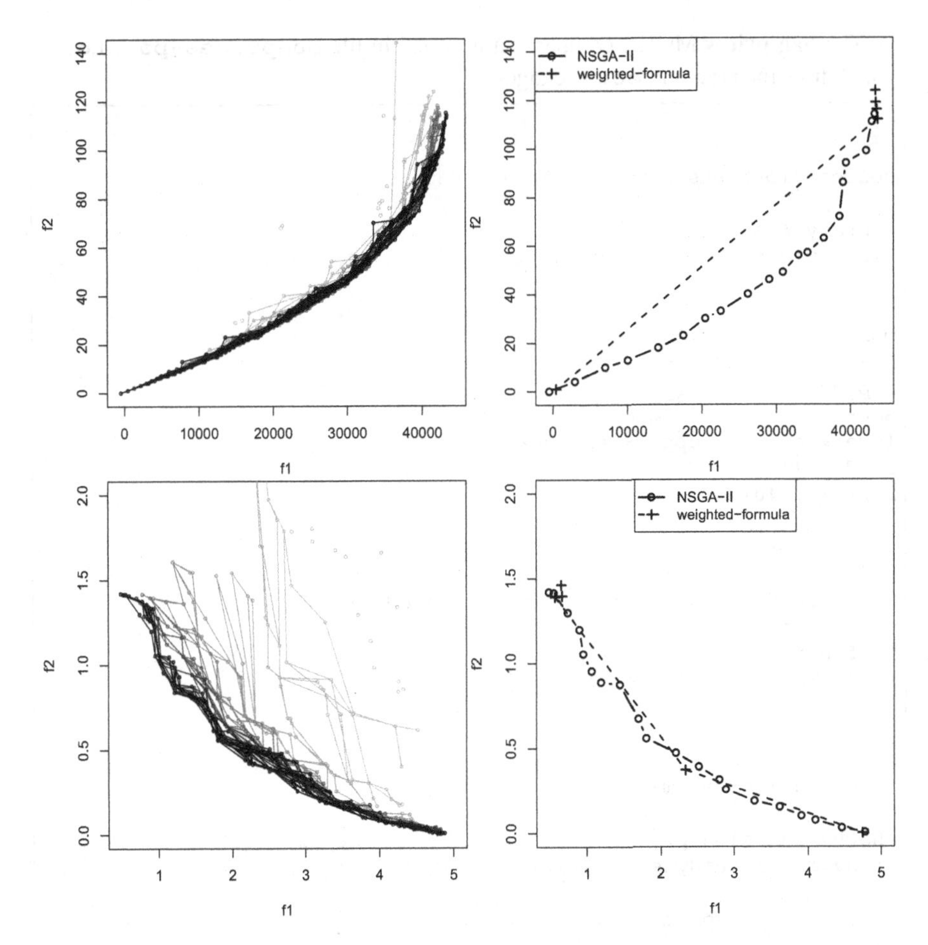

Fig. 6.3 NSGA-II results for **bag prices** (top graphs) and **FES1** (bottom graphs) tasks (left graphs show the Pareto front evolution, while right graphs compare the best Pareto front with the weighted-formula results)

a reference point is needed. The AS-EMOA requires expert domain knowledge to define an aspiration set (which includes several points) (Rudolph et al., 2014). Then, the aspiration set is used to create an approximation Pareto front that is close to the target aspiration set when assuming a Hausdorff distance, which measures the longest distance between two sets. NSGA-III is a variant of the NSGA-II method (Deb and Jain, 2014). The main difference is set in the selection operator, which creates and continuously updates a number of well-spread reference points. When compared with NSGA-II, NSGA-III is more suited for Many-Objective Optimization tasks, a term that is used when there is a higher number of objectives ($m > 3$). Both the **`MaOEA`** NSGA-III and **`ecr`** NSGA-II implementations employ a default Simulated Binary Crossover (SBX) crossover (Deb et al., 2007) and polynomial mutation (Liagkouras and Metaxiotis, 2013) to create new solutions.

The **bag prices** MOEA comparison is coded in file **bagprices-pareto.R**, which uses the **mco** and **ecr** packages:

```
### bagprices-pareto.R file ###

source("mo-tasks.R") # load multi-optimization tasks

library(mco) # load mco::nsga2
library(ecr) # load ecr::nsga2, smsemoa, asemoa

# --- integer bag prices task (common setup):
m=2 # two objectives
D=5 # 5 bag prices
Pop=20 # population size
Gen=100 # maximum number of generations (stop criterion)
lower=rep(1,D);upper=rep(1000,D)
Seed=12345
ref=c(0,140) # reference point: used by the hypervolume

# auxiliary functions --------------------------------
# ieval: integer evaluation (minimization goal):
ieval=function(x) c(-profit(x),produced(x)) # -f1,f2
# function that converts -f1 to f1:
f1=function(m) # m is a matrix
{ m[,1]= -1*m[,1]; return(m) } # f1 = -1 * -f1

# function that sorts m according to 1st column
o1=function(m) # m is a matrix
  return(m[order(m[,1]),])
# function that shows the hypervolume
showhv=function(p) # p is a Pareto front, uses global ref
 { hv=dominatedHypervolume(as.matrix(p),ref) # compute
      hypervolume
   cat("hypervolume=",hv,"\n")
 }
# -----------------------------------------------------

methods=c("mco::nsga2","ecr::nsga2","ecr::smsemoa","ecr:asemoa",
          "reference","aspiration")
cat("run ",methods[1],"\n")
set.seed(Seed)
s1=mco::nsga2(fn=ieval,idim=5,odim=m,
         lower.bounds=lower,upper.bounds=upper, # bounds
         popsize=Pop,generations=Gen)
p1=s1$value[s1$pareto.optimal,]
showhv(p1)
p1=o1(f1(p1)) # Pareto front: f1,f2

cat("run ",methods[2],"\n")
set.seed(Seed)
s2=ecr::nsga2(fitness.fun=ieval,minimize=TRUE,n.objectives=m,
              n.dim=D,lower=lower,upper=upper,mu=Pop,
              terminators=list(stopOnIters(Gen)))
p2=s2$pareto.front # convert to matrix
```

```
showhv(p2)
p2=o1(f1(p2)) # Pareto front: f1,f2

cat("run ",methods[3],"\n")
set.seed(Seed)
s3=smsemoa(fitness.fun=ieval,minimize=TRUE,n.objectives=m,
           n.dim=D,
           lower=lower,upper=upper,mu=Pop,ref.point=ref,
           terminators=list(stopOnIters(Gen)))
p3=s3$pareto.front # convert to matrix
showhv(p3)
p3=o1(f1(p3)) # Pareto front: f1,f2

cat("run ",methods[4],"\n")
set.seed(Seed)
# set the aspiration set (in this case, Pop points are used):
# basic aspiration with 2 lines and 1 inflection point
p0=matrix(ncol=Pop,nrow=m)
half1=round(Pop/2);half2=Pop-half1 # for 2 sequences
p0[1,]=c(seq(-5000,-30000,length.out=half1),
         seq(-30000,-45000,length.out=half2))
p0[2,]=c(seq(0,30,length.out=half1),
         seq(30,100,length.out=half2))
s4=asemoa(fitness.fun=ieval,minimize=TRUE,n.objectives=m,
          n.dim=D,aspiration=p0,
          lower=lower,upper=upper,mu=Pop,
          terminators=list(stopOnIters(Gen)))
p4=s4$pareto.front # convert to matrix
showhv(p4)
p4=o1(f1(p4)) # Pareto front: f1,f2

pdf("bagprices-pareto.pdf")
par(mar=c(4.0,4.0,0.1,0.1)) # set pdf margins
xlim=c(-1,45000);ylim=c(-1,141) # plot x and y limits
plot(p1,xlim=xlim,ylim=ylim,xlab="f1",ylab="f2",type="n")
col=c(paste("gray",round(seq(1,75,length.out=5)),sep=""),"black")
pch=c(1:3,5,4,18) # types of points used
# plot the 4 Pareto fronts:
lines(p1,type="b",cex=0.5,lwd=2,lty=1,pch=pch[1],col=col[1])
lines(p2,type="b",cex=0.5,lwd=2,lty=2,pch=pch[2],col=col[2])
lines(p3,type="b",cex=0.5,lwd=2,lty=3,pch=pch[3],col=col[3])
lines(p4,type="b",cex=0.5,lwd=2,lty=4,pch=pch[4],col=col[4])
# show reference point:
points(ref[1],ref[2],pch="X",cex=1.0,col=col[5])
abline(v=ref[1],col=col[5])
abline(h=ref[2],col=col[5])
# show aspiration set
p0=t(p0)
p0=f1(p0)
lines(p0,type="b",cex=1.0,lwd=1,lty=1,pch=pch[6],col=col[6])
legend("bottomright",methods,lwd=c(rep(2,4),1,1),
       pch=pch,lty=c(1:4,1,1),col=col)
dev.off()
```

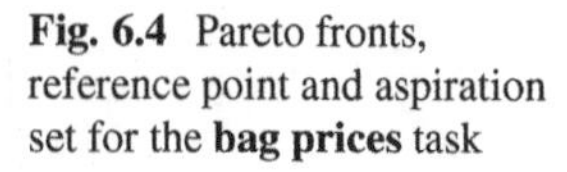

Fig. 6.4 Pareto fronts, reference point and aspiration set for the **bag prices** task

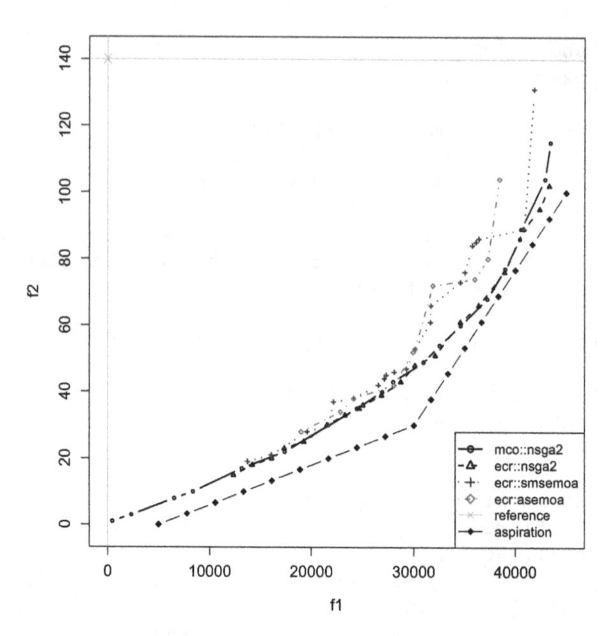

The file explores four MOEA algorithm implementations: **nsga2** (from **mco**, and **ecr**, NSGA-II); **smsemoa** (SMS-EMOA) and **asemoa** (AS-EMOA). The task setup includes the adopted values in the previous demonstration (e.g., a population size of **Pop=20**). For each algorithm execution, the hypervolume indicator is shown using reference point (0, 140), which was manually defined after looking at Fig. 6.3. Since AS-EMOA requires an aspiration set, a simple manually designed object was created (**p0**) and that consists of two segments joined in one inflection point. It should be noted that the **p0** object was created also looking at Fig. 6.3. Both the reference point and aspiration set can be visualized in Fig. 6.4. As for the hypervolume, the indicator is computed using the **mco** function **dominatedHypervolume**. The demonstration result is:

```
> source("bagprices-pareto.R")
run  mco::nsga2
hypervolume= 4346232
run  ecr::nsga2
hypervolume= 4347477
run  ecr::smsemoa
hypervolume= 4038532
run  ecr:asemoa
hypervolume= 3742906
```

In this execution, the best result was provided by the NSGA2 algorithm, with both **mco** and **ecr** implementations providing very similar and the highest hypervolume results. Figure 6.4 shows the distinct Pareto curves, confirming that for this execution the NSGA2 outperformed the SMS-EMOA and AS-EMOA algorithms.

While NSGA2 provided the best results for the **bag prices** task, it should be noted that only one run was executed and thus the comparison is not robust. A more fair comparison of multi-objective optimization algorithms is provided for the **fes1** demonstration, which uses the packages: **mco** (**dominatedHypervolume**), **ecr** (**nsga2** and **smsemoa**), **MaOEA** (**optimMaOEA**), and **tictoc** (**tic** and **toc**, to compute time elapsed).

Before detailing the demonstration, there are two relevant notes about the **MaOEA** package. First, the package has a high technical debt (Sect. 4.3), since it requires the installation of the Conda tool (https://docs.conda.io/) and several Python modules. Second, the explored NSGA-III implementation of **MaOEA** (version `0.5.2`) does not include lower and upper bound arguments. Thus, the initial population needs to be explicitly set with values within the [0, 1] range, which corresponds to the particular **fes1** task lower and upper values. However, if other tasks were addressed, with different real value bounds $x_i \in [L_i, U_i]$ and $i \in \{1, \ldots, D\}$, then the NSGA-III code would require an adaptation of the evaluation function. For instance, by rescaling first the solution ($x_i' = (U_i - L_i) + L_i$) and then evaluating the transformed solution.

The **fes1** comparison assumes a dimension of $D = 8$, a population size of $N_P = 20$ and $maxit = 100$ generations. Each algorithm is executed with 20 runs. The comparison analyses three types of results: full Pareto curves (one for each run), aggregated hypervolume values and aggregated Pareto curves. To aggregate the distinct run results, the Wilcoxon nonparametric test is used, which allows to compute a pseudo-median and its confidence interval. To compute a central Pareto curve, the procedure proposed in (Cortez et al., 2020) is adopted, which is applicable when $m = 2$ and performs a vertical (x-axis, f_1) aggregation. First, each curve is transformed, using a linear interpolation, to contain the same set of x-axis equally spaced points (according to fixed number of samples). Then, the vertical points (y-axis values) are grouped for the same x-axis point, allowing to compute a central measure (the pseudo-median) and its associated confidence interval. If needed, the procedure can easily be adapted to perform a horizontal aggregation by using the same principles (e.g., usage of y-axis equally spaced points).

The MOEA comparison using multiple runs is implemented in file **fes1-pareto.R**:

```
### fes1-pareto.R file ###

source("mo-tasks.R") # load multi-optimization tasks

library(mco) # load dominatedHypervolume
library(ecr) # load nsga2, smsemoa
# comment this line if you cannot install MaOEA
library(MaOEA) # load optimMaOEA
library(tictoc) # load tic and toc

# --- FES1 real value task (common setup):
m=2 # two objectives
D=8 # dimension
```

```
Pop=20 # population size
Gen=100 # maximum number of generations (stop criterion)
lower=rep(0,D);upper=rep(1,D) # lower and upper
Seed=12345 # random seed
ref=c(2.5,1.5) # reference point: used by the hypervolume
runs=20 # total number of execution runs

# auxiliary functions --------------------------------
# function that sorts m according to 1st column
o1=function(m) # m is a matrix
  return(m[order(m[,1]),])

# wilcox estimated median and confidence intervals:
wmedian=function(x,level=0.95) # x is a vector
{ # ignore wilcox.test warnings if they appear
  wx=suppressWarnings(wilcox.test(x,conf.level=level,
      conf.int=TRUE,alternative="two.sided",correct=TRUE))
  # use median if constant data in x:
  if(is.nan(wx$estimate)) wx$estimate=median(x)
  wx$conf.int[is.nan(wx$conf.int)]=c(wx$estimate,wx$estimate)
  lint=wx$estimate-wx$conf.int[1] # low conf. int. value
  uint=wx$conf.int[2]-wx$estimate # upper c.i. value
  return(c(wx$estimate,lint,uint)) # median,lower c.i.,upper c.i.
}

# create a vertical averaging curve with vsamples on x-axis:
#   Pareto is a vector list with runs results
#   xlim and ylim are the plot x-axis and y-axis limits
#   samples is the number of x-axis samples
vmedian=function(curves,xlim,ylim,samples=15)
{
 # target x-axis equally spaced range:
 xout=seq(xlim[1],xlim[2],length.out=samples)
 m=matrix(nrow=length(curves),ncol=samples)

 for(i in 1:length(curves)) # cycle all curves
 {
  # interpolate each curve to have points at xout values
  # (assumes min f1 and min f2 goals)
  acurve=suppressWarnings(approx(curves[[i]][,1],
          curves[[i]][,2],
          xout=xout,yleft=ylim[2],yright=ylim[1]))
  m[i,]=acurve$y # update the curve
 }
 vmed=matrix(nrow=4,ncol=samples)
 vmed[1,]=xout # first row
 # fill other rows with median,lower conf.int,upper conf.int
 for(j in 1:samples) # vertical average
 {
   vmed[2:4,j]=wmedian(m[,j])
 }
 return(vmed) # vertical median curve
}
```

```
# plot a vertical confidence interval bar (from file compare.R):
#   x are the x-axis points
#   ylower and yupper are the lower and upper y-axis points
#   ... means other optional plot parameters (lty, etc.)
confbar=function(x,ylower,yupper,K=100,...)
{ segments(x-K,yupper,x+K,...)
  segments(x-K,ylower,x+K,...)
  segments(x,ylower,x,yupper,...)
}

methods=c("nsga2","smsemoa","nsga3")
# methods=c("nsga2","smsemoa","nsga3") # uncomment if no MaOEA
nm=length(methods) # shorter variable

# wrapper function that runs a MOEA, returns a Pareto front
moeatest=function(method,fn,m,D,Pop,ref,lower,upper,Gen)
{
 if(method=="nsga2")
  { # NSGA-II
   s=ecr::nsga2(fitness.fun=fn,minimize=TRUE,n.objectives=m,
                n.dim=D,lower=lower,upper=upper,mu=Pop,
                terminators=list(stopOnIters(Gen)))
   # convert to a f1 sorted matrix:
   return(as.matrix(o1(s$pareto.front)))
  }
 else if(method=="smsemoa")
  { # SMS-EMOA
   s=smsemoa(fitness.fun=fn,minimize=TRUE,n.objectives=m,n.dim=D,
          lower=lower,upper=upper,mu=Pop,ref.point=ref,
          terminators=list(stopOnIters(Gen)))
   # convert to a f1 sorted matrix:
   return(as.matrix(o1(s$pareto.front)))
  }
 # comment this "else if" block if MaOEA is not installed -
 else if(method=="nsga3") # SBX and polynomial mutation
  { # NSGA-III
   # define the initial random population:
   pop=matrix(runif(Pop*D,min=lower[1],max=upper[1]),nrow=D)
   # control list NSGAIII defaults: pcx=1.0, pmut=1/D
   C=list(crossoverProbability=1.0,mutationProbability=1/D)
   s=optimMaOEA(x=pop,fun=fes1,solver=NSGA3,nObjective=m,
                nGeneration=Gen,seed=Seed,control=C)
   # convert to a f1 sorted matrix:
   return(as.matrix(o1(t(s$y))))
  }
 # end of #else if" block --------------------------------
}

# execute several runs of a MOEA optimization:
runtest=function(runs,Seed,method,fn,m,D,Pop,ref,lower,upper,Gen)
{
 set.seed(Seed) # set for replicability
 s=sample(1:Seed,runs) # seed for each run
 res=vector("list",runs)
```

```
 tic()
 for(i in 1:runs)
 {
  set.seed(s[i]) # seed for the run
  Pareto=moeatest(method,fn,m,D,Pop,ref,lower,upper,Gen)
  res[[i]]=Pareto
 }
 toc()
 return(res)
}

res=vector("list",runs) # store all results in res
hv=matrix(nrow=nm,ncol=runs) # hypervolume results
for(i in 1:nm)
{
 cat("execute ",runs,"runs with ",methods[i],"...\n")
 res[[i]]=runtest(runs,Seed,methods[i],fes1,m,D,Pop,ref,lower,
                 upper,Gen)
 # store all hypervolume run results for methods[i]:
 hv[i,]=unlist(lapply(res[[i]],dominatedHypervolume,ref))
 # show hypervolume median and confidence intervals:
 wm=round(wmedian(hv[i,]),2) # median and confidence intervals
 cat("median hypervolume: ",wm[1]," +- (",
     wm[2],",",wm[3],")\n",sep="")
}

xlim=c(0.7,5) # set manually for fes1
ylim=c(0,1.5) # set manually for fes1

# create pdf with all Pareto runs for all methods
pdf("fes1-all.pdf")
par(mar=c(4.0,4.0,0.1,0.1))
# set an empty plot frame:
plot(res[[1]][[1]],xlim=xlim,ylim=ylim,type="n",
     xlab="f1",ylab="f2")
col=paste("gray",round(seq(1,50,length.out=nm)))
#col=c("black","blue","red") # if you prefer color
lwd=c(1,1,2)
for(i in 1:nm) # cycle methods
{
 for(r in 1:runs) # cycle runs
 { # plot each Pareto curve for each run
  lines(res[[i]][[r]],type="l",col=col[i],lty=i,lwd=lwd[i])
 }
}
legend("topright",methods,col=col,lty=1:nm,lwd=lwd)
dev.off()

# create pdf to compare Pareto front curves:
#   method proposed in (Cortez et al., 2020)
pdf("fes1-median.pdf")
par(mar=c(4.0,4.0,0.1,0.1))
# set an empty plot frame:
plot(res[[1]][[1]],xlim=xlim,ylim=ylim,type="n",
```

```
    xlab="f1",ylab="f2")
samples=15
pch=1:nm
lwd=c(2,2,2)
K=diff(range(xlim))/samples/4 # 1/4 of the x-axis sample spacing
for(i in 1:nm)
{
 V=vmedian(res[[i]],xlim,ylim,samples=samples)
 # if needed clean artificially generated points (in extremes)
 if(i==2) ix=6:(samples-1) # set manually
 else if(i==3) ix=1:6 # set manually
 else ix=1:samples # all x-axis samples
 lines(V[1,ix],V[2,ix],type="b",pch=pch[i],col=col[i],lwd=lwd[i])
 #plotH(V[1,ix],V[2,ix],V[3,ix],V[4,ix],col=col[i])
 confbar(V[1,ix],V[2,ix]-V[3,ix],V[2,ix]+V[4,ix],K=K,col=col[i])
}
legend("topright",methods,col=col,lty=1,lwd=lwd,pch=pch)
dev.off()
```

The Wilcoxon estimates are provided by the **wilcox.test** function. Since this function can issue warnings and return **NaN** values (e.g., if the analyzed object includes constant values), the wrapper function **wmedian** was built to deal with these issues, returning a 3 element vector (with median, lower and upper confidence interval values). The demonstration assumes the standard 95% confidence level when computing the Wilcoxon statistics. Another auxiliary implemented function is **vmedian**, which computes the vertical averaging of the Pareto curves. The **vmedian** code uses the **approx** R function, which performs a linear interpolation of a set of data points. The interpolation is useful because each Pareto curve can have a different number of points, spread through a different x-axis range. The effect of the approx function is shown using two examples:

```
> approx(x=1:4,y=c(1,NA,NA,4),xout=0:4/2)
$x
[1] 0.0 0.5 1.0 1.5 2.0

$y
[1]  NA  NA 1.0 1.5 2.0

> approx(x=1:4,y=c(1,NA,NA,4),xout=0:4/2,yleft=0)
$x
[1] 0.0 0.5 1.0 1.5 2.0

$y
[1] 0.0 0.0 1.0 1.5 2.0
```

By using **approx**, all processed Pareto curves will have the same set of x-axis points (defined in **xout**). Then, the **vmedian** function computes the aggregated statistics for the processed curves. The demonstration code compares three MOEA algorithms: NSGA-II (**nsga2**), SMS-EMOA (**smsemoa**), and NSGA-III (**optimMaOEA**). The algorithms are executed with several runs in function **runtest**, which cycles the same wrapper function **moeatest** to execute each algorithm (defined by the **method** argument). The vector list **res** is used to store all Pareto curve results (for the three methods and **runs=20**). Then, the aggregated hypervolume indicator is computed and shown in the console. Finally, two PDF plots are produced: **fes1-all.pdf**, with all Pareto curves; and **fes1-median.pdf**, with the vertical median Pareto curves. To facilitate the comparison, the artificially produced points (due to the interpolation into a non-existing left or right x-axis Pareto region) were manually removed (e.g., **ix=1:7**) from the second PDF file.

The comparison demonstration assumes: the reference point **ref=c(2.5,1.5)**; **xlim=c(0.7,5)** and **ylim=c(0,1.5)** (for the x-axis and y-axis bounds); and **sample=15** x-axis points for the vertical aggregated curves. The resulting console output is:

```
> source("fes1-pareto.R")
execute  20 runs with  nsga2 ...
4.171 sec elapsed
median hypervolume: 1.15 +- (0.04,0.05)
execute  20 runs with  smsemoa ...
2.635 sec elapsed
median hypervolume: 0.05 +- (0.03,0.07)
execute  20 runs with  nsga3 ...
13.774 sec elapsed
median hypervolume: 1.23 +- (0.07,0.06)
```

In this demonstration, the tested NGSA-III algorithm requires more computational effort but it provides the best hypervolume results (median of 1.23, with the confidence interval [1.16, 1.29]). The second best hypervolume result is given by NSGA-II (median of 1.15). The obtained PDF plots are shown in Fig. 6.5. The plots confirm the worst performance that is obtained by SMS-EMOA. Regarding the NSGA-III and NSGA-II comparison, when analyzing all curves (left of Fig. 6.5), the former algorithm obtains better results for the left x-axis (f_1) range (e.g., $\in [0.7, 1.5]$). However, after a given x-axis range value (e.g., 1.5), the Pareto curves overlap and it is difficult to distinguish any differences. The vertical aggregated plot provides a simpler visual comparison and it is particularly helpful when the number of runs is high. Indeed, the right graph of Fig. 6.5 produces a more clear comparison, confirming that NSGA-III evolves a more interesting Pareto curve (with statistical confidence) for the low x-axis range ($f_1 \in [0.7, 1.5]$), while NSGA-II produces better Pareto values (statistically significant when compared with SMS-EMOA) for higher x-axis values ($f_1 \in [2.5, 5]$).

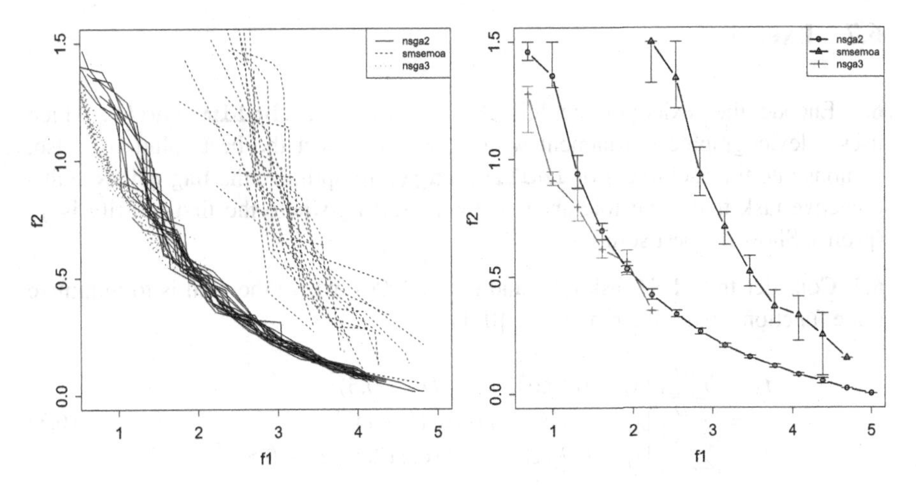

Fig. 6.5 Pareto curves for the **fes1** task: all curves (left) and vertically aggregated (right)

6.6 Command Summary

MaOEA	package with several many-objective algorithms
belegundu()	multi-objective belegundu test problem (package **mco**)
ecr	evolutionary computation in R package
ecr()	evolutionary computation function (package **ecr**)
is.matrix()	returns true if argument is a matrix
lrbga.bin()	lexicographic genetic algorithm (chapter file **"lg-ga.R"**)
mco	package for multi-criteria optimization algorithms
nsga2()	NSGA-II algorithm (package **mco**)
optimMaOEA()	run a many-objective algorithm (package **MaOEA**)
order()	rearrange a R object
paretoFilter()	extract the Pareto optimal front (package **mco**)
paretoSet()	returns the Pareto front from a **mco** result object (package **mco**)
tic()	start a timer (package **tictoc**)
tictoc	package for timing R scripts
toc()	compute time elapsed (package **tictoc**)
tournament()	tournament under a lexicographic approach (chapter file **"lg-ga.R"**)

6.7 Exercises

6.1 Encode the lexicographic hill climbing function `lhclimbing()`, which uses a lexicographic tournament with $k = 2$ to select the best solutions. Also, demonstrate the usefulness of `lhclimbing()` to optimize the **bag prices** multi-objective task under the tolerance vector $(0.1, 0.1)$, where the first priority is f_1 (profit). Show the best solution.

6.2 Consider the **FES2** task (Huband et al., 2006), where the goal is to minimize three functions under the range $x_i \in [0, 1]$:

$$\begin{array}{l} f_1 = \sum_{i=1}^{D} (x_i - 0.5\cos(10\pi i/D) - 0.5)^2 \\ f_2 = \sum_{i=1}^{D} |x_i - \sin^2(i-1)\cos^2(i-1)|^{0.5} \\ f_2 = \sum_{i=1}^{D} |x_i - 0.25\cos(i-1)\cos(2i-2) - 0.5|^{0.5} \end{array} \tag{6.3}$$

Use the hypervolume indicator, with a reference point of $(2.0, 8.0, 10.0)$, to compare the final Pareto front solutions when exploring two multi-optimization approaches:

1. a weight-based genetic algorithm (WBGA), which encodes a different weight vector into each solution of the genetic population (use function **`rgba`**); and
2. NSGA-II algorithm (via **`mco`** package).

For both approaches, use a population size of 20, 100 generations and a dimension of $D = 8$.

Chapter 7
Applications

7.1 Introduction

Previous chapters have approached demonstrative optimization tasks that were synthetically generated. The intention was to present a tutorial perspective and thus more simpler tasks were approached. As a complement, this chapter addresses real-world applications, where the data is obtained from a physical phenomenon. Exemplifying the optimization of real-world data in R is interesting for three main reasons. First, it demonstrates the value and potential impact of the modern optimization methods. Second, physical phenomena may contain surprising or unknown features. Third, it provides additional code examples of how to load and process real-world data.

This chapter addresses three real-world tasks that are discussed in the next sections: traveling salesman problem, time series forecasting, and wine quality classification.

7.2 Traveling Salesman Problem

The Traveling Salesman Problem (TSP) is a classical combinatorial optimization. The goal is to find the cheapest way of visiting all cities (just once) of a given map, starting in one city and ending in the same city (Reinelt, 1994). The complete traveling is known as Hamiltonian cycle or TSP tour. The standard TSP assumes symmetric costs, where the cost of visiting from city A to B is the same as visiting from B to A. This problem is non-deterministic polynomial (NP)-complete, which means that there is no algorithm capable of reaching the optimum solution in a polynomial time (in respect to n) for all possible TSP instances of size n (the total number of cities considered). The complex TSP task has been heavily studied in the last decades and a substantial improvement has been reached. For example, in

P. Cortez, *Modern Optimization with R*, Use R!,
https://doi.org/10.1007/978-3-030-72819-9_7

1954 an instance with 49 cities was the maximum TSP size solved, while in 2004 the record was established in 24,978 cities (Applegate et al., 2011).

Due to its importance, several optimization methods were devised to specifically address the TSP, such as 2-opt and concorde. The former method is a local search algorithm that is tested in this section and starts with a random tour and then exchanges two cities in order to avoid any two crossing paths until no improvements are possible (Croes, 1958). The latter is a more recent advanced exact TSP solver for symmetric TSPs and that is based on branch-and-cut approach (Applegate et al., 2001). More recently, Nagata and Kobayashi (2013) have proposed a state-of-the-art genetic algorithm that is capable of handling TSP instances with 200,000 cities and that provided competitive results when compared with a specialized TSP method, the Lin-Kernighan (LK) algorithm. The proposed method uses an initial greedy local search, based on the 2-opt method, to set the initial population. Then, a genetic algorithm is applied. This algorithm does not use mutations and new solutions are created by adopting the Edge Assembly Crossover (EAX), which is considered very effective for TSP. Rather than competing with the state-of-the art TSP methods, in this section the TSP is used as an application domain to show how ordered (or permutation based) representations can be handled by metaheuristics. Also, it is used as example to demonstrate how Lamarckian evolution works (2-opt is adopted to improve the solutions).

An ordered representation is a natural choice for TSP since it assumes that solutions can be generated as permutations of the symbols from an alphabet with n elements (the search dimension is $D = n$). Without losing generality, the alphabet can be defined by the integers in the set $\{1, 2, \ldots, n\}$, where each integer represents a city (e.g., Boston $\rightarrow$ 1, New York $\rightarrow$ 2) (Rocha et al., 2001). The TSP solution is thus represented by the sequence $(c_1, c_2, \ldots, c_n)$, where $c_i \in \{1, 2, \ldots, n\}$ represents a city and $c_i \neq c_j$ for any $i \neq j$ value. Under this representation, the search space consists of permutations of the alphabet, resulting in $n!$ possible solutions for a TSP instance of size n. For instance, for $n = 3$, the search space is (1,2,3), (1,3,2), (2,1,3), (2,3,1), (3,1,2), and (3,2,1). The adaption of modern optimization methods to this representation type requires assuring that generated solutions (e.g., created in the *initialization* and *change* functions of Algorithm 1) are feasible, avoiding missing integers and repeated values.

For single-state methods, several mutation operators can be adopted to change solutions, such as (Michalewicz and Fogel, 2004; Bossek, 2017): exchange, insertion, displacement, and scramble. The first operator swaps two randomly selected cities; the second operator inserts a city into a random position; the third operator inserts a random subtour into another position; and the fourth mutation rearranges a random subtour. Fig. 7.1 shows examples of these order mutations.

For evolutionary approaches, there are several crossover methods that preserve order, such as Partially Matched Crossover (PMX), Order Crossover (OX), and Cycle Crossover (CX) (Rocha et al., 2001). PMX first selects two cutting points and the corresponding matching section is exchanged between the two parents, through position-to-position exchange operations. The OX also exchanges two sections between the parents but keeping an emphasis on the relative order of the genes

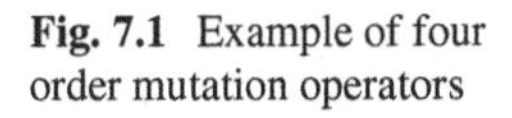

Fig. 7.1 Example of four order mutation operators

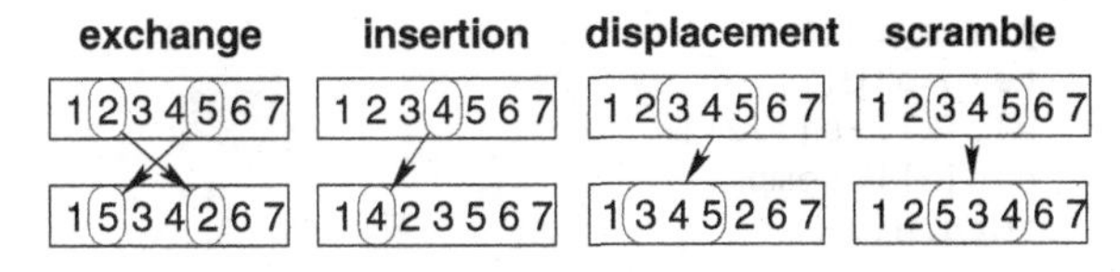

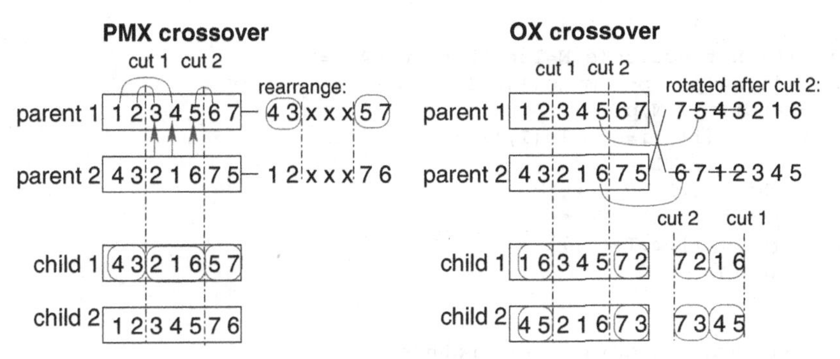

Fig. 7.2 Example of PMX and OX crossover operators

from both parents. Both operators are shown in Fig. 7.2, while CX is described in Exercise 7.1.

In this section, two modern optimization methods are adopted for the TSP task: simulated annealing and evolutionary algorithm (under three variants). Simulated annealing uses a mutation operator to change solutions, while the evolutionary variants use crossover and mutation operators. The evolutionary methods include: a simple permutation based evolutionary algorithm (EA); similarly to what is executed in (Nagata and Kobayashi, 2013), an initial greedy 2-opt search to set the initial population, followed by the simple evolutionary algorithm (EA2); and a Lamarckian evolution (LEA). The LEA method employs the EA method except that all evaluated solutions are first improved using the 2-opt procedure and then the improved solutions are updated into the LEA population. The **ecr** package is adopted to implement the evolutionary approaches, since it is a very flexible package, allowing to use several functions (e.g., for selection and breeding operators) and customize the main evolution cycle of the algorithm. The R code that implements the permutation operators and adapted evolutionary algorithm is presented in file **perm.R**:

```
### perm.R file ###

#-- functions to handle permutation optimization tasks --

library(ecr)

# operators for permutation solutions:

### mutation operators:
exchange=function(s,N=length(s))
```

```
{ p=sample(1:N,2) # select two positions
  temp=s[p[1]] # swap values
  s[p[1]]=s[p[2]]
  s[p[2]]=temp
  return(s)
}

insertion=function(s,N=length(s),p=NA,i=NA)
{ if(is.na(p)) p=sample(1:N,1) # select a position
  I=setdiff(1:N,p) # ALL except p
  if(is.na(i)) i=sample(I,1) # select random place
  if(i>p) i=i+1 # need to produce a change
  I1=which(I<i) # first part
  I2=which(I>=i) # last part
  s=s[c(I[I1],p,I[I2])] # new solution
  return(s)
}

displacement=function(s,N=length(s))
{ p=c(1,N)
  # select random tour different than s
  while(p[1]==1&&p[2]==N) p=sort(sample(1:N,2))
  I=setdiff(1:N,p[1]:p[2]) # ALL except p
  i=sample(I,1) # select random place
  I1=which(I<i) # first part
  I2=which(I>=i) # last part
  s=s[c(I[I1],p[1]:p[2],I[I2])]
  return(s)
}

scramble=function(s,N=length(s))
{ p=c(1,N)
  # select random tour different than s
  while(p[1]==1&&p[2]==N) p=sort(sample(1:N,2))
  # scramble p
  scr=sample(p[1]:p[2],p[2]-p[1]+1)
  I=setdiff(1:N,p[1]:p[2]) # ALL except p
  I1=which(I<p[1]) # first part
  I2=which(I>p[2]) # last part
  s=s[c(I[I1],scr,I[I2])]
  return(s)
}

### crossover operators:
# partially matched crossover (PMX) operator:
# m is a matrix with 2 parent x ordered solutions
pmx=function(m)
{ N=ncol(m)
  p=sample(1:N,2) # two cutting points
  c=m # children
  for(i in p[1]:p[2])
   { # rearrange:
     c[1,which(c[1,]==m[2,i])]=c[1,i]
     # crossed section:
```

```
      c[1,i]=m[2,i]
      # rearrange:
      c[2,which(c[2,]==m[1,i])]=c[2,i]
      # crossed section:
      c[2,i]=m[1,i]
    }
  return(c)
}

# order crossover (OX) operator:
# m is a matrix with 2 parent x ordered solutions
ox=function(m)
{ N=ncol(m)
  p=sort(sample(1:N,2)) # two cutting points
  c=matrix(rep(NA,N*2),ncol=N)
  # keep selected section:
  c[,p[1]:p[2]]=m[,p[1]:p[2]]
  # rotate after cut 2 (p[2]):
  I=((p[2]+1):(p[2]+N))
  I=ifelse(I<=N,I,I-N)
  a=m[,I]
  # fill remaining genes:
  a1=setdiff(a[2,],c[1,p[1]:p[2]])
  a2=setdiff(a[1,],c[2,p[1]:p[2]])
  I2=setdiff(I,p[1]:p[2])
  c[,I2]=rbind(a1,a2)
  return(c)
}

# auxiliary functions that:
# return same argument OR
#  population or fitness fields (if Lamarck evaluation)
pop_eF=function(pop,fitness)
{ if(!is.list(fitness)) return (pop)
  else # is.list (Lamarck)
   {
    for(i in 1:length(pop))
       pop[[i]]=fitness[,i]$sol
    return(pop)
   }
}
fit_eF=function(fitness)
{
  if(!is.list(fitness)) return (fitness)
  else # is.list (Lamarck)
   {
    res=matrix(ncol=ncol(fitness),nrow=1)
    for(i in 1:ncol(fitness)) res[,i]=fitness[,i]$eval
    return(res)
   }
}

# evolutionary algorithm with Lamarckian optimization
#  uses the same arguments as ecr():
```

```
# mu -- population size (if numeric)
#    or initial population (if vector list)
# perm -- the maximum integer size of the permutation
# trace -- if >0, then the best is shown every trace iter.
# fitness.fun might perform a Lamarckian change, if it
#   does, then it should return a list with $eval and $sol
eal=function(fitness.fun,mu,lambda,perm=perm,
             p.recomb=0.7,p.mut=0.3,n.elite=0L,maxit,trace=maxit)
{
 # control object: fitness.fun,mutation,crossover,selection
 control=initECRControl(fitness.fun=fitness.fun,n.objectives=1)
 control=registerECROperator(control,"mutate",mutInsertion)
 control=registerECROperator(control,"recombine",recOX)
 # use roulette wheel for selection of individuals and parents:
 control=registerECROperator(control,"selectForSurvival",
                             selTournament)
 control=registerECROperator(control,"selectForMating",
                             selTournament)

 if(is.numeric(mu)) pop=genPerm(mu,perm) # initial population
 else { pop=mu;mu=length(mu) } # use mu as initial population
 eF=evaluateFitness(control,pop) # evaluation
 pop=pop_eF(pop,eF) # update (if Lamarck)
 fitness=fit_eF(eF) # update (if Lamarck), matrix (1 x mu)

 if(trace>0) cat("initial pop. best:",min(fitness),"\n")
 for (i in seq_len(maxit)) # main evolutionary cycle
   {
    # sample lambda individuals:
    idx=sample(1:mu,lambda)
    # create and evaluate offspring:
    fitidx=matrix(fitness[,idx],ncol=lambda)
    offspring=generateOffspring(control,pop[idx],fitness=fitidx,
                               lambda=lambda,p.recomb=0.7,p.mut
                                   =0.3)
    fitness.o=evaluateFitness(control,offspring)
    offspring=pop_eF(offspring,fitness.o) # update (if Lamarck)
    fitness.o=fit_eF(fitness.o) # get (if Lamarck)
    # selection of best solutions:
    sel=replaceMuCommaLambda(control,pop,offspring,fitness,
                             fitness.o,n.elite=n.elite)
    pop=sel$population # update population
    fitness=sel$fitness # update fitness
    if(i%%trace==0) cat("gen:",i,"best:",min(fitness),"\n")
   }
 # pop is a vector list of size mu, fit is a matrix (1 x mu)
 return(list(pop=pop,fit=fitness))
}
```

File **perm.R** implements all mentioned permutation operations. Also, it defines the new evolutionary algorithm with an option for Lamarckian evolution in **eal()**. This algorithm uses solutions that are represented as permutations, with the

$\{1, 2, \ldots, n\}$ alphabet, where **perm=**n. Most of the algorithm code are based on **ecr** functions:

- **initECRControl**—create a control;
- **registerECROperator**—update control with an operator;
- **genPerm**—create a random population with permutations;
- **evaluateFitness**—evaluate a population;
- **seq_len**—generate a regular sequence;
- **generateOffspring()**—apply crossover and mutation; and
- **replaceMuCommaLambda**—select the next population.

The algorithm is customized to include: an insertion mutation, an OX crossover, tournament selection (for both parent and population selection). The algorithm accepts the **mu** argument, which defines the size of the population to be randomly generated (if numeric) or the initial population (if a vector list). Then, the initial population is evaluated. Since the fitness function can change the individuals (if Lamarckian evolution is used), the auxiliary **pop_eF** and **fit_eF** were implemented to update the respective objects. Next, the main evolution cycle is executed, which corresponds to a typical evolutionary algorithm. In each iteration, **lamda** offspring individuals are generated and evaluated. Then, the next generation is defined, by selecting the best individuals (using tournament selection) from the current and offspring populations. Finally, the cycle ends after **maxit** generations and the last population is returned.

An important aspect of **eal** is the option to perform a Lamarckian evolution scheme (explained in Sect. 1.6). The option is controlled by the evaluation function (**fitness.fun**). If a single evaluation numeric value is returned, then the standard evolutionary algorithm is assumed. Else a list should be returned with two fields: **$eval**—the fitness value; and **$sol**—the original or an improved solution. When a list is returned, then it is assumed that the population individual will be changed after its evaluation (the Lamarckian variant).

To demonstrate the simulated annealing and evolutionary algorithms, the Qatar TSP was selected and that includes 194 cities. Other national TSPs are available at http://www.math.uwaterloo.ca/tsp/world/countries.html. The R code is presented in file **tsp.R**:

```
### tsp.R file ###

library(TSP) # load TSP package
library(RCurl) # load RCurl package
library(ecr) # load ecr package
source("perm.R") # load permutation operators

# get Qatar - 194 cities TSP instance:
txt=getURL("http://www.math.uwaterloo.ca/tsp/world/qa194.tsp")
# simple parse of txt object, removing header and last line:
txt=strsplit(txt,"NODE_COORD_SECTION") # split text into 2 parts
txt=txt[[1]][2] # get second text part
txt=strsplit(txt,"EOF") # split text into 2 parts
```

```
txt=txt[[1]][1] # get first text part
# save data into a simple .csv file, sep=" ":
cat(txt,file="qa194.csv")
# read the TSP format into Data
# (first row is empty, thus header=TRUE)
# get city Cartesian coordinates

Data=read.table("qa194.csv",sep=" ")
Data=Data[,3:2] # longitude and latitude
names(Data)=c("x","y") # x and y labels
 # number of cities

# distance between two cities (EUC_2D-norm)
# Euclidean distance rounded to whole number
D=dist(Data,upper=TRUE)
D[1:length(D)]=round(D[1:length(D)])
# create TSP object from D:
TD=TSP(D)

set.seed(12345) # for replicability
cat("2-opt run:\n")
PTM=proc.time() # start clock
R1=solve_TSP(TD,method="2-opt")
sec=(proc.time()-PTM)[3] # get seconds elapsed
print(R1) # show optimum
cat("time elapsed:",sec,"\n")

MAXIT=50000 # maximum number of evaluations
Methods=c("SANN","EA","EA2","LEA") # comparison of 4 methods

RES=matrix(nrow=MAXIT,ncol=length(Methods))
MD=as.matrix(D)
N=nrow(MD)

# overall distance of a tour (evaluation function):
tour=function(s) # s is a solution (permutation)
{ # compute tour length:
  EV<<-EV+1 # increase evaluations
  s=c(s,s[1]) # start city is also end city
  res=0
  for(i in 2:length(s)) res=res+MD[s[i],s[i-1]]
  # store in global memory the best values:
  if(res<BEST) BEST<<-res
  if(EV<=MAXIT) F[EV]<<-BEST
  # only for hybrid method:
  # return tour
  return(res)
}

# Lamarckian evaluation function:
#  uses 2-opt to improve solution s, returns a list with:
#  eval - the tour for improved solution
#  sol  - the improved solution
ltour2opt=function(s)
```

```
{
 EV<<-EV+1 # increase evaluations
 # improve s using "2-opt" method:
 s2=solve_TSP(TD,method="2-opt",control=list(tour=s))
 res=attr(s2,"tour_length") # solve_TSP computes the tour
 if(res<BEST) BEST<<-res
 if(EV<=MAXIT) F[EV]<<-BEST
 return( list(eval=res,sol=as.numeric(s2)) )
}

Trace=1000 # show at the console the optimization evolution

cat("SANN run:\n")
set.seed(12345) # for replicability
s=sample(1:N,N) # initial solution
EV=0; BEST=Inf; F=rep(NA,MAXIT) # reset these vars.
C=list(maxit=MAXIT,temp=2000,trace=TRUE,REPORT=Trace)
PTM=proc.time() # start clock
SANN=optim(s,fn=tour,gr=insertion,method="SANN",control=C)
sec=(proc.time()-PTM)[3] # get seconds elapsed
cat("time elapsed:",sec,"best tour:",F[MAXIT],"\n")
RES[,1]=F

# EA: simple permutation based evolutionary algorithm
cat("EA run:\n")
set.seed(12345) # for replicability
EV=0; BEST=Inf; F=rep(NA,MAXIT) # reset these vars.
popSize=30;lambda=round(popSize/2);Eli=1
maxit=ceiling((MAXIT-popSize)/(lambda-1))

PTM=proc.time() # start clock
EA=eal(fitness.fun=tour,mu=popSize,lambda=lambda,
       perm=N,n.elite=Eli, # elitism
       maxit=maxit,trace=Trace)
sec=(proc.time()-PTM)[3] # get seconds elapsed
cat("time elapsed:",sec,"best tour:",F[MAXIT],"\n")
RES[,2]=F

# EA2: initial 2-opt search + EA
cat("EA2 run:\n")
set.seed(12345) # for replicability
EV=0; BEST=Inf; F=rep(NA,MAXIT) # reset these vars.
maxit=ceiling((MAXIT-popSize)/(lambda-1))
PTM=proc.time() # start clock
# apply 2-opt local search to initial population:
pop=genPerm(popSize,N) # create random population
control=initECRControl(fitness.fun=ltour2opt,n.objectives=1)
eF=evaluateFitness(control,pop)
pop=pop_eF(pop,eF) # update population
# run standard evolutionary algorithm
EA2=eal(fitness.fun=tour,mu=pop,lambda=lambda,
       perm=N,n.elite=Eli, # elitism
       maxit=maxit,trace=Trace)
sec=(proc.time()-PTM)[3] # get seconds elapsed
```

```
cat("time elapsed:",sec,"best tour:",F[MAXIT],"\n")
RES[,3]=F

cat("LEA run:\n")
popSize=30;lambda=round(popSize/2);Eli=1
set.seed(12345) # for replicability
EV=0; BEST=Inf; F=rep(NA,MAXIT) # reset these vars.
maxit=ceiling((MAXIT-popSize)/(lambda-1))
PTM=proc.time() # start clock
LEA=eal(fitness.fun=ltour2opt,mu=popSize,lambda=lambda,
        perm=N,n.elite=Eli, # elitism
        maxit=maxit,trace=Trace)
sec=(proc.time()-PTM)[3] # get seconds elapsed
cat("time elapsed:",sec,"best tour:",F[MAXIT],"\n")
RES[,4]=F

# create PDF with comparison:
pdf("qa194-opt.pdf",paper="special")
par(mar=c(4.0,4.0,0.1,0.1))
X=seq(1,MAXIT,length.out=200)
ylim=c(min(RES)-50,max(RES))
plot(X,RES[X,1],ylim=ylim,type="l",lty=4,lwd=2,
     xlab="evaluations",ylab="tour distance")
lines(X,RES[X,2],type="l",lty=3,lwd=2)
lines(X,RES[X,3],type="l",lty=2,lwd=2)
lines(X,RES[X,4],type="l",lty=1,lwd=2)
legend("topright",Methods,lwd=2,lty=4:1)
dev.off()

# create 3 PDF files with best tours:
pdf("qa194-2-opt.pdf",paper="special")
par(mar=c(0.0,0.0,0.0,0.0))
plot(Data[c(R1[1:N],R1[1]),],type="l",xaxt="n",yaxt="n")
dev.off()
pdf("qa194-ea.pdf",paper="special")
par(mar=c(0.0,0.0,0.0,0.0))
b=EA$pop[[which.min(EA$fit)]]
plot(Data[c(b,b[1]),],type="l",xaxt="n",yaxt="n")
dev.off()
pdf("qa194-lea.pdf",paper="special")
par(mar=c(0.0,0.0,0.0,0.0))
b=LEA$pop[[which.min(LEA$fit)]]
plot(Data[c(b,b[1]),],type="l",xaxt="n",yaxt="n")
dev.off()
```

The code starts by reading the Qatar TSP instance from the Web by using the **getURL** function of the **RCurl** package. The data is originally in the TSPLIB Format (extension **.tsp**) and thus some parsing (e.g., remove the header part until **NODE_COORD_SECTION**) is necessary to convert it into a CSV format. The national TSPs assume a traveling cost that is defined by the Euclidean distance rounded to the nearest whole number (TSPLIB EUC_2D-norm). This is easily computed by using the R **dist** function, which returns a distance matrix between all rows of a data matrix.

The code tests five methods to solve the Qatar instance: 2-opt, simulated annealing, EA, EA2, and LEA. The first method is executed using the **TSP** package, which is specifically addressed to handle the TSP and includes two useful functions: **TSP**—generates a TSP object from a distance matrix; and **solve_TSP**—solves a TSP instance under several **method** options (e.g., **"2-opt"** and **"concorde"**). To simplify the code and analysis, the remaining optimization methods are only compared under a single run, although a proper comparison would require the use of several runs, as shown in Sect. 4.5. The simulated annealing and evolutionary algorithms are executed under the same conditions. Similarly to the code presented in Sect. 4.5, the global **EV**, **BEST**, and **F** are used to trace the evolution of optimization according to the number of function evaluations. The method parameters were fixed into a temperature of $T = 2000$ for the simulated annealing and population size of $L_P = 30$ for the evolutionary algorithm (with **lambda**=$L_P/2$).

The **tour()** evaluation function uses an already defined distance matrix (**MD** object) to save computational effort (i.e., the Euclidean distance is only calculated once). The **ltour2opt** evaluates and returns the solution improved by the 2-opt search. This function uses the R function **attr**, which gets or sets an attribute from an object (in this case, it is used to get the **tour_length** attribute). The same initialization seed is used by the metaheuristics, which are monitored up to **MAXIT=50000** function evaluations. The simulated annealing uses the insertion operator to change solutions, while the evolutionary methods use two operators (insertion and OX). For each method, the code shows the length of the tour and time elapsed (in seconds). The code also generates three PDF files with a comparison of simulated annealing and evolutionary approaches and two optimized tours (for 2-opt, EA and LEA). The execution result of file **tsp.R** is:

```
> source("tsp.R")
2-opt run:
object of class TOUR
result of method 2-opt for 194 cities
tour length: 10289
time elapsed: 0.025
SANN run:
sann objective function values
initial       value 90785.000000
iter    10000 value 47065.000000
iter    20000 value 44832.000000
iter    30000 value 42634.000000
iter    40000 value 40280.000000
iter    49999 value 40280.000000
final         value 40280.000000
sann stopped after 49999 iterations
time elapsed: 14.096 best tour: 40280
EA run:
initial pop. best: 87938
gen: 1000 best: 30086
gen: 2000 best: 22657
```

```
gen: 3000 best: 21293
time elapsed: 26.769 best tour: 21196
EA2 run:
initial pop. best: 10004
gen: 1000 best: 9976
gen: 2000 best: 9949
gen: 3000 best: 9930
time elapsed: 26.353 best tour: 9930
LEA run:
initial pop. best: 10004
gen: 1000 best: 9395
gen: 2000 best: 9395
gen: 3000 best: 9395
time elapsed: 112.479 best tour: 9352
```

The specialized 2-opt approach is much faster than the metaheuristics, requiring just 0.03 s. The most computationally demanding method is the Lamarckian variant (LEA), which requires around 4 times more computation (112.48 s) when compared with the normal EA (26.77 s). The extra computation is explained by the intensive use of the 2-opt search, given that all evaluations require the call to the **solve_TSP** function.

The comparison between the simulated annealing (SANN) and evolutionary methods is shown in Fig. 7.3. Under the experimental setup conditions, the simulated annealing initially performs similarly to the standard evolutionary algorithm. However, after around 3000 evaluations the simulated annealing improvement gets slower when compared with the standard evolutionary algorithm and after around 35,000 evaluations, the convergence is rather flat, reaching a tour value of 40,280. The pure evolutionary algorithm (EA) outperforms the simulated annealing and after 50,000 evaluations it obtains a final tour of 21,196. More importantly, both EA2 and LEA perform much better than SANN and EA methods, denoting clear differences from the first evaluations (Fig. 7.3). EA2 obtained a final tour value of 9930, while LEA achieved the best TSP result (9352). A visual comparison of the EA, 2-opt, and LEA methods is shown in Fig. 7.4, revealing a clear improvement in the quality of the solutions when moving from left to right.

In this demonstration, the Lamarckian approach provided the best results. This method is based on the 2-opt, which also is used in EA2 (to set the initial population). The 2-opt procedure was specifically proposed for the symmetrical and standard TSP and thus it performs a more clever use of the distance matrix (domain knowledge) to solve the task. As explained in Sect. 1.1, the simulated annealing and evolutionary algorithms are general-purpose methods that are often used in a "black box" fashion, with an indirect access to the domain knowledge through the received evaluation function values. This black box usage means that the metaheuristics do not take advantage of specific TSP knowledge, leading to worst results. Yet, the advantage is that the same algorithms could be easily adapted (by adjusting the evaluation function) to other TSP variants (e.g., with constrains) or combinatorial problems (e.g., job shop scheduling) while 2-opt (or even this particular Lamarckian method) could not.

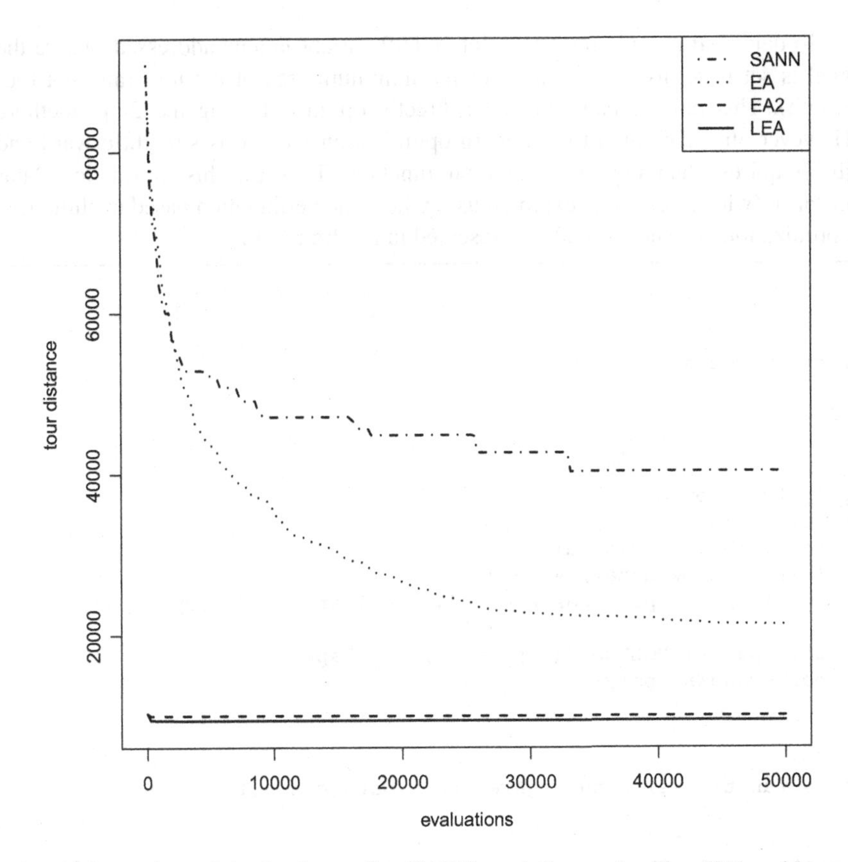

Fig. 7.3 Comparison of simulated annealing (SANN), evolutionary algorithm (EA), and Lamarckian approach (LEA) approaches for the Qatar TSP

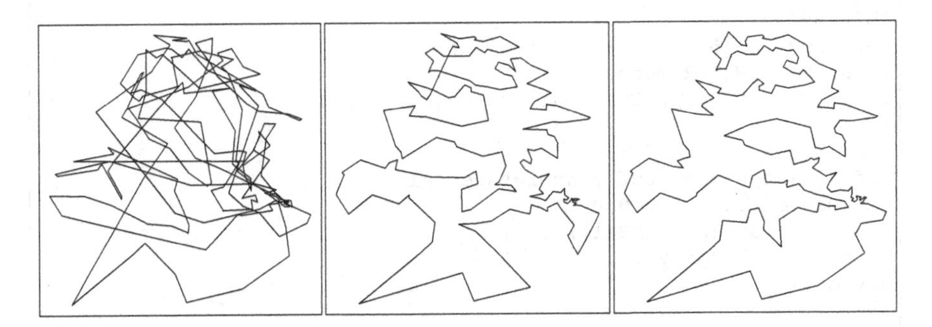

Fig. 7.4 Optimized tour obtained using evolutionary algorithm (left), 2-opt (middle), and Lamarckian evolution (right) approaches for the Qatar TSP

To demonstrate the previous point, a TSP variant is now addressed, where the goal is set in terms of searching for the minimum area of the tour (and not tour length). This new variant cannot be directly optimized using the 2-opt method. However, the adaptation to a modern optimization method is straightforward and just requires changing the evaluation function. To show this, a smaller Qatar instance (with 25 cities) is explored using the same permutation based evolutionary optimization. The new R code is presented in file **tsp2.R**:

```
### tsp2.R file ###
# this file assumes that tsp.R has already been executed

library(rgeos) # get gArea function

N2=25 # lets consider just 25 cities

# creates a polygon (rgeos package) object from TSP data
poly=function(data)
{ poly="";sep=", "
  for(i in 1:nrow(data))
  { if(i==nrow(data)) sep=""
    poly=paste(poly,paste(data[i,],collapse=" "),sep,sep="")
  }
  poly=paste("POLYGON((",poly,"))",collapse="")
  poly=readWKT(poly) # WKT format to polygon
}

# new evaluation function: area of polygon
area=function(s) return(gArea(poly(Data[c(s,s[1]),])))

# new data with N2 cities:
Data2=Data[1:N2,]
D2=dist(Data2,upper=TRUE)
D2[1:length(D2)]=round(D2[1:length(D2)])
# create TSP object from D:
TD2=TSP(D2)
set.seed(12345) # for replicability
R2=solve_TSP(TD2,method="2-opt")
cat("area of 2-opt TSP tour:",area(R2),"\n")

# plot area of 2-opt:
pdf("qa-2opt-area.pdf",paper="special")
par(mar=c(0.0,0.0,0.0,0.0))
PR1=poly(Data[c(R2,R2[1]),])
plot(PR1,col="gray")
dev.off()

# EA:
cat("EA run for TSP area:\n")
set.seed(12345) # for replicability
pSize=30;lambda=round(pSize/2);maxit=40;Eli=1
PTM=proc.time() # start clock
OEA=eal(fitness.fun=area,mu=popSize,lambda=lambda,
        perm=N2,n.elite=Eli, # elitism
```

```
        maxit=maxit,trace=maxit)
sec=(proc.time()-PTM)[3] # get seconds elapsed
bi=which.min(OEA$fit)
b=OEA$pop[[bi]]
cat("best fitness:",OEA$fit[1,bi],"time elapsed:",sec,"\n")

# plot area of EA best solution:
pdf("qa-ea-area.pdf",paper="special")
par(mar=c(0.0,0.0,0.0,0.0))
PEA=poly(Data[c(b,b[1]),])
plot(PEA,col="gray")
lines(Data[c(b,b[1]),],lwd=2)
dev.off()
```

The evaluation function uses the **gArea()** function of the **rgeos** package to compute the area of a polygon. Before calculating the area, the function first converts the selected solution into a polygon object by calling the **poly** auxiliary function. The latter function first encodes the tour under the Well Known Text (WKT) format (see http://en.wikipedia.org/wiki/Well-known_text) and then uses **readWKT()** (from the **rgeos** package) function to create the polygon (**sp geometry** object used by the **rgeos** package). For comparison purposes, the area is first computed for the tour optimized by the 2-opt method. Then, the evolutionary optimization is executed and stopped after 40 iterations. The code also produces two PDF files with area plots related to the best solutions optimized by the evolutionary algorithm and 2-opt methods.

The result of executing file **tsp2.R** is:

```
> source("tsp2.R")
area of 2-opt TSP tour: 232744.7
EA run for TSP area:
initial pop. best: 5659.719
gen: 40 best: 267.8175
best fitness: 267.8175 time elapsed: 1.447
```

Now the evolutionary approach achieves a value (267.8) that is much better when compared with 2-opt (232,744.7). The area of each optimized solution is shown in Fig. 7.5. As shown by the plots, the evolutionary algorithm best solution presents several crossing paths, which reduces the tour area. In contrast, 2-opt intentionally avoids crossing paths and such strategy is very good for reducing the path length but not the tour area, thus resulting in a worst area value.

7.3 Time Series Forecasting

A univariate time series is a collection of timely ordered observations related with an event ($y_1, y_2, \ldots, y_t$) and the goal of time series forecasting (TSF) is to model a complex system as a black box, predicting its behavior based in historical data (Makridakis et al., 1998). Past values (called in-samples) are first used to fit the

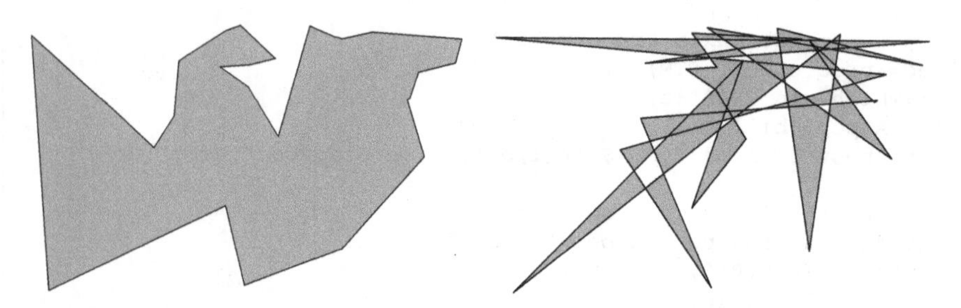

Fig. 7.5 Area of the 25 city Qatar instance tour given by 2-opt (left) and optimized by the evolutionary approach (right)

model and then forecasts are estimated ($\hat{y}_t$) for future values (called out-of-samples). The TSF task is to determine a function f such that $\hat{y}_t = f(y_{t-1}, y_{t-2}, \ldots, y_{t-k})$, where k denotes the maximum time lag used by the model. Under one-step ahead forecasting, the errors (or residuals) are given by $e_i = y_i - \hat{y}_i$, where $i \in \{T + 1, T + 2, \ldots, T + h\}$, T is the current time, h is the horizon (or ahead predictions), and the errors are to be minimized according to an accuracy metric, such as the mean absolute error ($MAE = \frac{\sum_i |e_i|}{h}$) (Stepnicka et al., 2013). TSF is highly relevant in distinct domains, such as Agriculture, Finance, Sales, and Production. TSF forecasts can be used to support individual and organizational decision making (e.g., for setting early production plans).

Due to its importance, several statistical TSF methods were proposed, such as the autoregressive integrated moving-average (ARIMA) methodology, which was proposed in 1976 and is widely used in practice (Makridakis et al., 1998). The methodology assumes three main steps: model identification, parameter estimation, and model validation. The ARIMA base model is set in terms of a linear combination of past values (AR component of order p) and errors (MA component of order q). The definition assumed by the R **arima** function is

$$\hat{x}_t = a_1 x_{t_1} + \ldots + a_p x_{t_p} + e_t + b_1 e_{t_1} + \ldots + b_q e_{t_q} \tag{7.1}$$

where $x_t = y_t - m$ and $m, a_1, \ldots, a_p$ and $b_1, \ldots, b_q$ are coefficients that are estimated using an optimization method. The **arima()** default is to use a conditional sum of squares search to find starting values and then apply a maximum likelihood optimization. To identify and estimate the best ARIMA model, the **auto.arima** function is adopted from the **forecast** package.

In this section, genetic programming (**rgb** package; if needed, Sect. 5.10 explains how to install the package) and grammatical evolution (**gramEvol**) are adopted to fit a time series by using a variable sized numeric function expression that includes: arithmetic operators (**+**, **-**, ***** and **/**), numeric constants, and two past time lags (y_{t-1} and y_{t-2}). As explained in Sects. 5.10 and 5.11, these evolutionary search methods present the advantage of finding explicit solutions that are easy to be interpreted by humans. The selected series is the sunspots numbers (also known as Wolf number), which measures the yearly number of dark spots present at the surface of the sun. Forecasting this series is relevant due to several reasons (Kaboudan, 2003): the sunspots data generation process is unknown; sunspots are often used to estimate solar activity levels; accurate prediction of sunspot numbers is a key issue for weather forecasting and for making decisions about satellite orbits and space missions. The data range from 1700 to 2019. In this demonstration, sunspots data from the years 1700–1987 are used as in-samples and the test period is 1988–2019 (out-of-samples). Forecasting accuracy is measured using the MAE metric and 1-step ahead forecasts ($h = 1$). The respective R code is presented in file **tsf.R**:

```
### tsf.R file ###

# get sunspot yearly series:
url="http://www.sidc.be/silso/DATA/SN_y_tot_V2.0.txt"
series=read.table(url)
# lets consider data from the 1700-2019 years:
# lets consider column 2: sunspot numbers
series=series[1:320,2]
# save to CSV file, for posterior usage (if needed)
write.table(series,file="sunspots.csv",
            col.names=FALSE,row.names=FALSE)

L=length(series) # series length
forecasts=32 # number of 1-ahead forecasts
outsamples=series[(L-forecasts+1):L] # out-of-samples
sunspots=series[1:(L-forecasts)] # in-samples

# mean absolute error of residuals
maeres=function(residuals) mean(abs(residuals))

# fit best ARIMA model:
INIT=10 # initialization period (no error computed before)
library(forecast) # load forecast package
arima=auto.arima(sunspots) # detected order is AR=2, MA=1
print(arima) # show ARIMA model
cat("arima fit MAE=",
    maeres(arima$residuals[INIT:length(sunspots)]),"\n")

# one-step ahead forecasts:
# (this code is needed because forecast function
#  only performs h-ahead forecasts)
LIN=length(sunspots) # length of in-samples
f1=rep(NA,forecasts)
for(h in 1:forecasts)
  { # arima with fixed coefficients and more in-samples:
   arima1=arima(series[1:(LIN+h-1)],
           order=arima$arma[c(1,3,2)],fixed=arima$coef)
   f1[h]=forecast(arima1,h=1)$mean[1]
  }
e1=maeres(outsamples-f1)
text1=paste("arima (MAE=",round(e1,digits=1),")",sep="")
```

```
### evolutionary methods:
# evaluate time series function
#   receives f: a function (GP) or expression (GE)
#   if(h>0) then returns 1-ahead forecasts
#   else returns MAE over fitting period (in-samples)
evalts=function(f,h=0) # fitness function
{ # global object previously defined: series,LIN
  if(h>0) y=series else y=series[1:LIN]
  LTS=length(y)
  F=rep(0,LTS) # forecasts
  E=rep(0,LTS) # residuals
  if(h>0) I=(LTS-h+1):LTS # h forecasts
  else I=INIT:LTS # fit to in-samples
  for(i in I)
    {
     if(class(f)=="function") F[i]=f(y[i-1],y[i-2])
     else F[i]=eval(f) # expr includes: y[i-1] and y[i-2]
     if(is.nan(F[i])) F[i]=0 # deal with NaN
     E[i]=y[i]-F[i]
    }
  if(h>0) return (F[I]) # forecasts
  else return(maeres(E[I])) # MAE on fit
}

Pop=200; Maxit=500; Eli=1 # common setup

# fit genetic programming (GP) arithmetic model:
RGP=require(rgp) # load rgp
if(RGP){ # this code is only executed if rgp is installed
 ST=inputVariableSet("y1","y2") # order of AR component
 cF1=constantFactorySet(function() rnorm(1)) # mean=0, sd=1
 FS=functionSet("+","*","-","/") # arithmetic
 # GP time series function
 #   receives function f
 #   if(h>0) then returns 1-ahead forecasts
 #   else returns MAE over fitting period (in-samples)
 gpmut=function(func) # GP mutation function
 { mutateSubtree(func,funcset=FS,inset=ST,conset=cF1,
                 mutatesubtreeprob=0.3,maxsubtreedepth=4)}
 set.seed(12345) # set for replicability
 # run the GP:
 cat("run GP:\n")
 gp=geneticProgramming(functionSet=FS,inputVariables=ST,
       constantSet=cF1,populationSize=Pop,eliteSize=Eli,
       fitnessFunction=evalts,
       stopCondition=makeStepsStopCondition(Maxit),
       mutationFunction=gpmut,verbose=TRUE)
 f2=evalts(gp$population[[which.min(gp$fitnessValues)]],
         h=forecasts)
 e2=maeres(outsamples-f2)
 text2=paste("gp (MAE=",round(e2,digits=1),")",sep="")
 cat("best solution:\n")
 print(gp$population[[which.min(gp$fitnessValues)]])
```

```
 cat("gp fit MAE=",min(gp$fitnessValues),"\n")
}

# fit grammatical evolution (GE) arithmetic model:
GE=require(gramEvol) #
if(GE) # this code is only executed if gramEvol is installed
{ # set the grammar rules:
 ruleDef=list(expr=gsrule("<expr><op><expr2>","<expr2>"),
              op=gsrule("+","-","*","/"),
              expr2=gsrule("y[i-1]","y[i-2]","<value>"),
              value=gsrule("<digits>.<digits>"),
              digits=gsrule("<digits><digit>","<digit>"),
              digit=grule(0,1,2,3,4,5,6,7,8,9)
             )
 gDef=CreateGrammar(ruleDef) # grammar object
 # simple monitoring function:
 monitor=function(results)
 { # print(str(results)) shows all results components
  iter=results$population$currentIteration # current iteration
  f=results$best$cost # best fitness value
  if(iter==1||iter%%100==0) # show 1st and every 100 iter
    cat("iter:",iter,"f:",f,"\n")
 }
 set.seed(12345) # set for replicability
 # run the GE:
 cat("run GE:\n")
 ge=GrammaticalEvolution(gDef,evalts,optimizer="es",
                         popSize=Pop,elitism=Eli,
                         iterations=Maxit,monitorFunc=monitor)
 b=ge$best # best solution
 cat("best solution:\n")
 print(b$expression)
 cat("ge fit MAE=",b$cost,"\n")
 f3=evalts(b$expressions,h=forecasts)
 e3=maeres(outsamples-f3)
 text3=paste("ge (MAE=",round(e3,digits=1),")",sep="")
}

# show quality of one-step ahead forecasts:
ymin=min(c(outsamples,f1))
if(RGP) ymin=min(ymin,f2)
if(GE) ymin=min(ymin,f3)
ymax=max(c(outsamples,f1))
if(RGP) ymax=max(ymax,f2)
if(GE) ymax=max(ymax,f3)
pdf("fsunspots.pdf")
par(mar=c(4.0,4.0,0.1,0.1))
plot(outsamples,ylim=c(ymin,ymax),type="b",pch=1,
     xlab="time (years after 1988)",ylab="values",cex=0.8)
lines(f1,lty=2,type="b",pch=2,cex=0.5)
pch=1:2;lty=1:2;text=c("sunspots",text1)
if(RGP)
 { lines(f2,lty=3,type="b",pch=3,cex=0.5)
   lty=c(lty,3);pch=c(pch,3)
```

```
    text=c(text,text2)
  }
if(GE)
  { lines(f3,lty=4,type="b",pch=4,cex=0.5)
    lty=c(lty,4);pch=c(pch,4)
    text=c(text,text3)
  }
legend("topright",text,lty=lty,pch=pch)
dev.off()
```

The ARIMA model is automatically found using the **auto.arima** function, which receives as inputs the in-samples. For this example, the identified model is an $ARIMA(2, 0, 1)$, with $p = 2$ and $q = 1$. The **forecast** function (from package **forecast**) executes multi-step ahead predictions. Thus, one-step ahead forecasts are built by using an iterative call to the function, where in each iteration the ARIMA model is computed with one extra in-sample value.

Regarding, the evolutionary methods, the **require** R base function is used to load the **rgp** and **gramEvol** packages. This function is very similar to the common **library** command, the main difference is that it returns a Boolean value (**TRUE** or **FALSE**) that corresponds to success in loading the package. This value is used by a **if** conditional structure, allowing to execute the genetic programming or grammatical evolution code only if their respective packages are installed. The evaluation function (**evalts**) is used by both evolutionary methods. It uses the same p order and thus the input variables are: y_{t-1}—**y1** (**rgp**) or **y[i-1]** (**gramEvol**); y_{t-2}—**y2** (**rgp**) or **y[i-2]** (**gramEvol**). In order to save code, the **evalts** function is used under two execution goals: fitness function, computing the MAE over all in-samples except for the first **INIT** values (when **h=0**); and estimation of forecasts, returning **h** one-step ahead forecasts. Since the **/** operator can generate **NaN** values (e.g., **0/0**), any **NaN** value is transformed into 0. To simplify the demonstration, only one run is used, with a fixed seed. The evolutionary methods are run using the same population size (**Pop=200**, elitism of **Eli=1**) and maximum number of generations (**Maxit=500**). The methods were configured using a setup similar to what was used in Sects. 5.10 and 5.11. In this demonstration, the evolutionary optimizer of the grammatical evolution was set to an evolution strategy (**optimizer="es"**). After finishing the algorithms execution, a PDF file is created, comparing the forecasts with the sunspots values. The result of executing file **tsf.R** is:

```
> source("tsf.R")
Series: sunspots
ARIMA(2,0,1) with non-zero mean

Coefficients:
         ar1      ar2      ma1     mean
      1.4652  -0.7587  -0.1599  78.7196
s.e.  0.0523   0.0472   0.0774   4.2936

sigma^2 estimated as 655.4:  log likelihood=-1341.73
AIC=2693.46   AICc=2693.68   BIC=2711.78
```

```
arima fit MAE= 19.29169
run GP:
STARTING genetic programming evolution run (Age/Fitness/
    Complexity Pareto GP search-heuristic) ...
evolution step 100, fitness evaluations: 9900, best fitness: 23
    .533647, time elapsed: 13.91 seconds
evolution step 200, fitness evaluations: 19900, best fitness: 19
    .319194, time elapsed: 31.61 seconds
evolution step 300, fitness evaluations: 29900, best fitness: 18
    .888153, time elapsed: 55.39 seconds
evolution step 400, fitness evaluations: 39900, best fitness: 18
    .846652, time elapsed: 1 minute, 19.8 seconds
evolution step 500, fitness evaluations: 49900, best fitness: 18
    .763177, time elapsed: 1 minute, 42.28 seconds
Genetic programming evolution run FINISHED after 500 evolution
    steps, 49900 fitness evaluations and 1 minute, 42.28 seconds.
best solution:
function (y1, y2)
y1 - (y1 - 0.280572562490903/(y1/y1 - 0.131997033458382/
    -0.762574707733437) *
    (y2 - y1))/(y1 + y2) * (y2 - (y1 - 0.280572562490903 * y2))
<bytecode: 0x7f9328e70e40>
gp fit MAE= 18.76318
run GE:
iter: 1 f: 28.38351
iter: 100 f: 23.98721
iter: 200 f: 23.98721
iter: 300 f: 23.98721
iter: 400 f: 23.98721
iter: 500 f: 23.98721
best solution:
expression(y[i - 1] - y[i - 2]/5.6)
ge fit MAE= 23.98721
```

The $ARIMA(2, 0, 1)$ model fits the in-samples with a MAE of 19.3. The genetic programming obtains a slightly better fitness ($MAE = 18.8$), while the grammatical evolution produces a worst fit ($MAE = 24.0$). It should be noted that the evolutionary solutions do not include the MA terms (i.e., past errors) of ARIMA. The evolved grammatical evolution expression is quite simple ($y_{i-1} - y_{i-2}/5.6$), while the best genetic programming solution is more complex and nonlinear (due to the / and * operators). The quality of the one-step ahead forecasts is shown in Fig. 7.6. All methods produce predictions that are close to the true sunspot values. Overall, the genetic programming solution produces slightly better forecasts with an improvement of 0.6 when compared with the ARIMA method in terms of MAE measured over the out-of-samples. This is an interesting result, since ARIMA methodology was specifically designed for TSF while genetic programming is a generic optimization method.

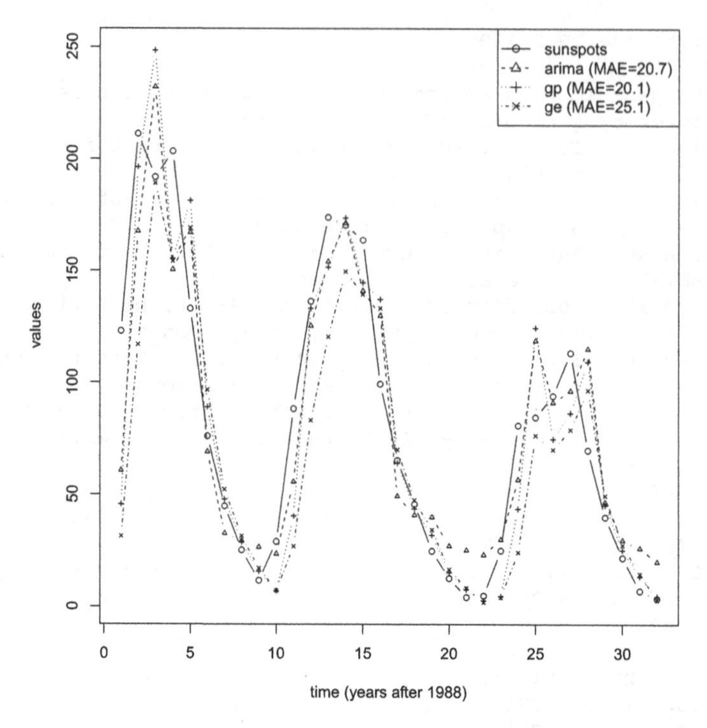

Fig. 7.6 Sunspot one-step ahead forecasts using ARIMA, genetic programming (gp) and grammatical evolution (ge) methods

7.4 Wine Quality Classification

Classification is an important data mining/machine learning task, where the goal is to build a data-driven model (i.e., model fit using a dataset) that is capable of predicting a class label (output target) given several input variables that characterize an item (Cortez, 2012). Typically, the dataset consists of tabular data (e.g., **data.frame**), where each row represents an example (or instance) and each column contains an attribute value. It is assumed that there is a categorical target attribute y, called output, that contains a known set of possible values (the class labels). There are also other attributes $(x_1, x_2, \ldots, x_I)$, known as inputs or features, that characterize or influence the phenomenon represented by the output attribute (y). The classification goal is to estimate an implicit function f that maps the I input values into the desired target y. This is achieved by using a training dataset (e.g., with past examples of the phenomenon) and a machine-learning algorithm. Using an iterative procedure, the algorithm searches for the best classification model (f, using a particular knowledge representation form defined by the algorithm) that optimizes an evaluation function that is computed using the training set (e.g., classification accuracy). Once the data-driven model is fit, it can be used to perform y predictions for new examples, provided that their x_i input values are known, where each prediction estimate is computed using: $\hat{y} = f(x_1, \ldots, x_I)$.

A simple example is the classification of the type of bank credit client, where $y \in$ {“good”,“bad”} (binary classification). Several input variables could be used to infer the target, such as the status of her/his bank account, credit purpose, and amount. The training set could be made of thousands of past records from a banking system, allowing to fit a machine-learning model. Then, the fitted model can be used to estimate the worthiness (“good” or “bad”) of new bank credit requests, thus supporting bank credit assignment decisions.

Often, it is possible to assign probabilities ($p \in [0, 1]$) for a class label when using a classifier. Under such scheme, the choice of a label is dependent of a decision threshold D, such that the class is true if $p > D$. The receiver operating characteristic (ROC) curve shows the performance of a two class classifier across the range of possible threshold (D) values, plotting one minus the specificity (false positive rate—FPR; x-axis) versus the sensitivity (also known as true positive rate—TPR; y-axis) (Fawcett, 2006). The overall accuracy is given by the Area Under the Curve ($AUC = \int_0^1 ROC dD$), measuring the degree of discrimination that can be obtained from a given model. The ROC analysis is a popular and richer measure for evaluating classification discrimination capability, particularly for binary classification tasks. The main advantage of the ROC curve is that performance is not dependent on the class labels distributions and error costs (Fawcett, 2006). Since there is a trade-off between specificity and sensitivity errors, the option for setting the best D value can be left for the domain expert. The ideal classifier should present an AUC of 1.0, while an AUC of 0.5 denotes a random classifier. Often, AUC values are read as (Gonçalves et al., 2020): 0.7—good; 0.8—very good; and 0.9—excellent.

Given the interest in classification, several machine-learning methods have been proposed, each one with its own purposes and advantages. In this section, the XGBoost model is adopted, which is an ensemble decision tree method proposed by Chen and Guestrin (2016) and that uses boosting to enhance the prediction results. This machine-learning algorithm provided competitive classification results in several Kaggle competitions (https://www.kaggle.com/dansbecker/xgboost). XGBoost contains several hyperparameters that need to be tuned in order to provide a better classification performance. These hyperparameters are often set by using trial-and-error experiments, typically by using a holdout or k-fold cross-validation method (Kohavi, 1995) to split the training data into training and validation sets. Using a fixed set of hyperparameters, the model is first fit using the training set and then its classification capability is measured on the (unseen) validation set, allowing to compute a performance measure (e.g., AUC). Classification performance is also affected by the input variables used to fit the model (this includes XGBoost and other machine-learning algorithms). Feature selection, i.e., the selection of the right set of input variables, is useful to discard irrelevant variables, leading to simpler models that are easier to interpret and often present a higher predictive accuracy (Guyon and Elisseeff, 2003).

For non-expert users, finding the best machine-learning configuration for a particular dataset is a nontrivial task. In effect, there is an increasing interest in the topic of Automated Machine Learning (AutoML), which performs an automatic search for the best data-driven machine-learning model (Hutter et al., 2019). This section approaches an AutoML task that is focused on a single algorithm (XGBoost) and that performs a simultaneous input feature and model hyperparameter selection, by adopting a multi-objective optimization search. As explained in (Freitas, 2004), the multi-objective strategy is justified by the trade-off that exists between having less features and increasing the classifier performance. The use of a modern optimization method, such as the NSGAII algorithm adopted in this section, is particularly appealing as an AutoML tool for non-specialized data science/machine-learning users, given that the search is fully automatic and more exhaustive, thus tending to provide better performances when compared with the manual classifier design.

The University California Irvine (UCI) machine-learning repository (Bache and Lichman, 2013) contains hundreds of datasets that are used by the data mining community to test algorithms and tools. Most datasets are related with classification tasks. In particular, this section explores the **wine quality** that was proposed and analyzed in (Cortez et al., 2009). The goal is to model wine quality based on physicochemical tests that are measured during the wine certification process. The output target (y = quality) was computed as the median of at least three sensory evaluations performed by wine experts, using a scale that ranges from 1 (very poor) to 10 (excellent). The physicochemical tests include $I = 11$ continuous variables (inputs), such as x_5 = chlorides and x_{11} = alcohol (vol.%). As explained in (Cortez et al., 2009), building a data-driven model, capable of predicting wine quality from physicochemical values, is important for the wine domain because the relationships between the laboratory tests and sensory analysis are complex and are still not fully understood. Thus, an accurate data-driven model can support the wine expert decisions, aiding the speed and quality of the evaluation performance. Also, such model could also be used to improve the training of oenology students.

This section exemplifies how the best classification model can be optimized by performing a simultaneous feature and model selection. The solution is thus represented the sequence $(b_1, b_2, \ldots, b_I, h_1, h_2, \ldots, h_{hmax})$, where $b_i \in \{0, 1\}$ is a binary number that defines if input variable x_i is used (true if b_i=1) and h_j denotes an hyperparameter value, whose nature depends of the selected machine-learning algorithm (e.g., integer or real value).

Given that two objectives are defined, improving classification performance and reducing the number of features used by the model, a multi-objective approach is adopted. The example code uses the **mco** and **rminer** packages. The former is used to get the **nsga2** function (NSGAII algorithm). The latter library facilitates the use of data mining algorithms in classification and regression tasks by presenting a short and coherent set of functions (Cortez, 2010). The **rminer** package is only briefly described here, for further details consult **help(package=rminer)**. The **rminer** package uses the **xgboost** function from the **xgboost** library to implement the XGBoost algorithm. For the wine quality classification task, the total number of inputs is $I = 11$. Also, $hmax = 4$ **xgboost** hyperparameters are considered. Based on https://www.hackerearth.com/practice/machine-learning/machine-learning-algorithms/beginners-tutorial-on-xgboost-parameter-tuning-r/tutorial/, the search ranges were defined as:

- h_1 =**nrounds**—integer $\in \{1,2,\ldots,200\}$;
- h_2 =**eta**—real value $\in [0.0,1.0]$;
- h_3 =**gamma**—real value $\in [0,10]$; and
- h_4 =**max_depth**—integer $\in \{0,1,\ldots,12\}$.

To simplify the R code, the **nsga2** solution representation assumes 15 $(11+4)$ real values $\in [0.0,1.0]$. The b_i values are obtained by rounding the real value genes to the {0, 1} range. As for the hyperparameters, a **decode** function is implemented, which transforms each original gene $\in [0,1]$ into its true search range by using a linear rescaling function (**transf**), where $x' = x(max - min) + min$ for the target search range $[min, max]$.

The classification example for the white wine quality dataset is based in two files:

- **wine-quality.R**—executes the full NSGAII search and saves the relevant obtained objects into files for their posterior usage; and
- **wineApp/app.R**—provides a **shiny** interactive Web application based on the previously obtained results (the **app.R** file needs to be inserted in a directory/folder called **wineApp**).

The demonstration file **wine-quality.R** is:

```
### wine-quality.R file ###

library(rminer) # load rminer package
library(mco) # load mco package

# load wine quality dataset directly from UCI repository:
file="http://archive.ics.uci.edu/ml/machine-learning-databases/
    wine-quality/winequality-white.csv"
d=read.table(file=file,sep=";",header=TRUE)

# convert the output variable into 2 classes of wine:
# "low" <- 3,4,5 or 6; "high" <- 7, 8 or 9
d$quality=cut(d$quality,c(1,6,10),c("low","high"))

# to speed up the demonstration, only 25% of the data is used:
n=nrow(d) # total number of samples
ns=round(n*0.25) # select a quarter of the samples
set.seed(12345) # for replicability
ALL=sample(1:n,ns) # contains 25% of the index samples
w=d[ALL,] # new wine quality data.frame
# show the attribute names:
cat("attributes:",names(w),"\n")
cat("output class distribution (25% samples):\n")
print(table(w$quality)) # show distribution of classes
```

```
# save dataset to a local CSV file:
write.table(w,"wq25.csv",col.names=TRUE,row.names=FALSE,sep=";")

# holdout split into training (70%) and test data (30%):
H=holdout(w$quality,ratio=0.7)
cat("nr. training samples:",length(H$tr),"\n")
cat("nr. test samples:",length(H$ts),"\n")
# save to file the holdout split index:
save(H,file="wine-H.txt",ascii=TRUE)

output=ncol(w) # output target index (last column)
maxinputs=output-1 # number of maximum inputs

# auxiliary functions:
# rescale x from [0,1] to [min,max] domain:
transf=function(x,min,max) return (x*(max-min)+min)
# decode the x genome into the model hyperparameters:
decode=function(x)
{
 # 4 xgboost hyperparameters for default "gbtree":
 nrounds=round(transf(x[1],1,200)) # [1,200]
 eta=x[2] # [0.0,1.0]
 gamma=transf(x[3],0,10) # [0,10]
 max_depth=round(transf(x[4],0,12)) # {0,...,12}
 return(c(nrounds,eta,gamma,max_depth))
}

# evaluation function (requires some computation):
eval=function(x)
{
 # read input features: from position 1 to maxinputs
 features=round(x[1:maxinputs]) # 0 or 1 vector
 inputs=which(features==1) # indexes with 1 values
 # use first feature if inputs is empty
 if(length(inputs)==0) inputs=1
 J=c(inputs,output) # attributes
 k3=c("kfold",3,123) # internal 3-fold validation
 # read hyperparameters:
 hpar=decode(x[(maxinputs+1):length(x)])
 M= suppressWarnings(try(
    mining(quality~.,w[H$tr,J],method=k3,
           model="xgboost",nrounds=hpar[1],
           eta=hpar[2],gamma=hpar[3],max_depth=hpar[4])
                      ,silent=TRUE))
 # AUC for the internal 3-fold cross-validation:
 if(class(M)=="try-error") auc=0.5 # worst auc
 else auc=as.numeric(mmetric(M,metric="AUC"))
 auc1=1-auc # maximization into minimization goal
 ninputs=length(inputs) # number of input features
 EVALS<<-EVALS+1 # update evaluations
 if(EVALS==1||EVALS%%Psize==0) # show current evaluation:
    cat(EVALS," evaluations (AUC: ",round(auc,2),
        " nr.features:",ninputs,")\n",sep="")
```

```
 return(c(auc1,ninputs)) # 1-auc,ninputs
}

# NSGAII multi-objective optimization:
cat("NSGAII optimization:\n")
m=2 # two objectives: AUC and number of input features
hxgb=4 # number of hyperparameters for xgboost
genome=maxinputs+hxgb # genome length
lower=rep(0,genome)
upper=rep(1,genome)
EVALS<<-0 # global variable
PTM=proc.time() # start clock
Psize=20 # population size
s1=mco::nsga2(fn=eval,idim=length(lower),odim=m,
        lower.bounds=lower,upper.bounds=upper,
        popsize=Psize,generations=10)
sec=(proc.time()-PTM)[3] # get seconds elapsed
cat("time elapsed:",sec,"\n")
# save to file the optimized Pareto front:
save(s1,file="wine-s1.txt",ascii=TRUE)
```

In this example, the **read.table** function is used to read the CSV file directly from the UCI repository. Originally, there are seven numeric values for the wine quality variable (range from 3 to 9). To simplify the demonstration execution times and visualization of the obtained results, some simplifications were assumed:

- Rather than classifying all 7 quality classes (from 3 to 9), a simpler binary task is adopted. Thus the **cut** R function is used to transform the numeric values into two classes ("low" and "high").
- The original dataset includes 4898 samples. To reduce the computational effort, this demonstration uses only 25% of the samples (randomly selected).
- Classification performance should be accessed over unseen data, not used for fitting. For a robust validation, typically an internal (to get a validation error to guide the search) and external 10-fold cross-validation (to measure the selected classifier generalization capabilities) should be adopted. However, this since this substantially increases the computational effort, a lighter approach was adopted. The internal validation is based on a 3-fold cross-validation, while the external validation is set in terms of a holdout split. The split assumes 70% of the selected samples for the search of the best model. The other 30% of the samples are used as test set, for estimating the model true generalization capabilities. The **holdout** function (from the **rminer** package) creates a list with the training (**$tr**) and test (**$ts**) indexes related with an output target.

The evaluation function is based on the powerful **mining** function (from **rminer**), which trains and tests a classifier under several runs and a given validation method. In this case, the used function arguments were:

- **x=quality~.**—a R **formula** means that the target is the quality attribute and all other **data** attributes (**~.**) are used as inputs;

- **data=w[H$tr,attributes]**—dataset used (**data.frame**), in this case corresponds to the training set samples (**H$tr**) and the data attributes defined by the solution **x** (selected **features** and **output**);
- **method=k3**—the estimation method is used by the function (in this case a 3-fold cross-validation initialized with the 123 seed);
- **model="xgboost"**—executes the **xgboost** fit function; and
- **eta=hpar[2], gamma=hpar[3], max_depth=hpar[4]**—sets the **xgboost** hyperparameter values.

For some search configurations (e.g., no input features), the **xgboost** function produces warnings and errors. Thus, the **suppressWarnings** and **try** R functions were used to handle such cases. At the end, the evaluation function returns the AUC value (computed using the **mmetric rminer** function) and the number of features. When there is an error (**"try-error"**) there is no classification model and thus the AUC is set to its worse possible value (0.5). The NSGAII algorithm is executed using a small population size (20) and stopped after 10 generations, in order to reduce the computational effort of the demonstration. To facilitate the visualization of the optimization results in the console, the global object **EVALS** is defined, allowing to track the number of elapsed evaluations. Every 20 evaluations, the evaluation function shows the AUC value and number of features of the evaluated individual. Several obtained results are saved into files, including the processed wine quality dataset (**w**), the holdout index (**H**), and Pareto front (**s1**), allowing their usage by the interactive demonstration. In case of **H** and **s1** objects, the saved files are created by using the **save** R function. The **wine-quality.R** console execution result is:

```
> source("wine-quality.R")
attributes: fixed.acidity volatile.acidity citric.acid
    residual.sugar chlorides free.sulfur.dioxide
    total.sulfur.dioxide density pH sulphates alcohol quality
output class distribution (25% samples):

 low high
 963  261
nr. training samples: 857
nr. test samples: 367
NSGAII optimization:
1 evaluations (AUC: 0.78 nr.features:7)
20 evaluations (AUC: 0.76 nr.features:4)
40 evaluations (AUC: 0.8 nr.features:6)
60 evaluations (AUC: 0.78 nr.features:6)
80 evaluations (AUC: 0.63 nr.features:3)
100 evaluations (AUC: 0.77 nr.features:3)
120 evaluations (AUC: 0.77 nr.features:2)
140 evaluations (AUC: 0.75 nr.features:4)
160 evaluations (AUC: 0.5 nr.features:1)
180 evaluations (AUC: 0.8 nr.features:7)
200 evaluations (AUC: 0.77 nr.features:2)
220 evaluations (AUC: 0.79 nr.features:3)
time elapsed: 43.27
```

The code starts by showing the dataset attribute names and output class distribution (25% of the data). The output classes are biased, where most of the samples (79%) are related to poor or average wines. Then, the console output shows the evolution of the NSGAII optimization, which requires a total of 43.3 s.

The second demonstration file (**wineApp/app.R**) can be executed in another R session (using the R console or RStudio IDE) and it provides an interactive Web visualization of the previously obtained NSGAII results:

```
### wineApp/app.R file ###

library(shiny) # load shiny
library(rminer) # load rminer

#---  global variables ------
# load the previously saved files
# (the files are in the upper level "../" directory):
w=read.table("../wq25.csv",header=TRUE,sep=";",
             stringsAsFactors=TRUE)
load("../wine-H.txt") # loads into H the holdout index
load("../wine-s1.txt") # loads into s1 the Pareto front

# global objects related with the data (w):
output=ncol(w) # output target index (last column)
maxinputs=output-1 # number of maximum inputs
Y=w[H$ts,"quality"] # target test data

# global objects related with the Pareto front (s1):
po=which(s1$pareto.optimal) # optimal points
NP=length(po) # number of Pareto front points
# sort Pareto front according to f2 (number of features):
i=sort.int(s1$value[po,2],index.return=TRUE)
pareto=s1$value[po[i$ix],] # Pareto front f1,f2 values
pop=s1$par[po[i$ix],] # Pareto solutions

# User Interface (UI):
ui=fluidPage( # begin fluidPage
   titlePanel("wineApp"), # title of the panel
   sidebarLayout( # sidebar with input and outputs
      position = c("right"), # put input at the right
   sidebarPanel(  # panel for input
    numericInput(
      inputId="number",
      label=paste("Pareto front point (1 to ",NP,"):",sep=""),
      min=1,  # minimum value
      max=NP, # maximum value
      step=1, # select only integers
      value=1 # default value
    )), # end sidebarPanel
   mainPanel( # panel for outputs
    h5("input features:"), # show fixed text
    verbatimTextOutput("text1"),
    h5("hyper parameters:"), # show fixed text
    verbatimTextOutput("text2"),
```

```
    splitLayout( # show horizontally 2 plots
      cellWidths = 350, # enlarge plot size
      plotOutput(outputId="par"),
      plotOutput(outputId="roc")
    ) # end splitLayout
   ) # end mainPanel
  ) # end sidebar
) # end fluidPage

# Server function:
server=function(input,output)
{
  # reactive component: return the selected model
  #  (only executed once for each input$number change)
  modelInput=reactive({ selectmodel(input$number) })
  # output components:
  output$text1=renderText({ # show inputs
    S=modelInput()
    paste(S$ninputs)
  })
  output$text2=renderText({ # show hyperparameters
    S=modelInput()
    paste(S$hpar)
  })
  output$par=renderPlot({ # plot Pareto curve
    plot(1-pareto[,1],pareto[,2],xlab="AUC",
         ylab="nr. features",type="b",lwd=2,
         main="Pareto curve:")
    points(1-pareto[input$number,1],pareto[input$number,2],
           pch="X",cex=1.5)
  })
  output$roc=renderPlot({ # plot ROC curve
    S=modelInput()
    mgraph(Y,S$P,graph="ROC",main=S$main,
           Grid=10,baseline=TRUE,leg="xgboost")
  })
}

# auxiliary functions:
# rescale x from [0,1] to [min,max] domain:
transf=function(x,min,max) return (x*(max-min)+min)
# decode the x genome into the model hyperparameters:
decode=function(x)
{
 # 4 xgboost hyperparameters for default "gbtree":
 nrounds=round(transf(x[1],1,200)) # [1,200]
 eta=x[2] # [0.0,1.0]
 gamma=transf(x[3],0,10) # [0,10]
 max_depth=round(transf(x[4],0,12)) # {0,...,12}
 return(c(nrounds,eta,gamma,max_depth))
}
# select from the Pareto front the classifier
#  i - index of the sorted Pareto front
selectmodel=function(i)
```

```
{
  x=pop[i,] # selected genome
  # decode the model:
  features=round(x[1:maxinputs])
  inputs=which(features==1)
  ninputs=names(w)[inputs]
  J=c(inputs,output) # data attributes
  hpar=decode(x[(maxinputs+1):length(x)])
  if(pareto[i,1]==0.5) # random classifier
     P=cbind(rep(1,length(Y)),rep(0,length(Y))) # predict "low"
  else
   {
    M=fit(quality~.,w[H$tr,J],model="xgboost",
          nrounds=hpar[1],eta=hpar[2],
          gamma=hpar[3],max_depth=hpar[4])
    P=predict(M,w[H$ts,J]) # get xgboost predictions
   }
  auc=mmetric(Y,P,metric="AUC") # compute the AUC
  main=paste("test data ROC curve"," (AUC=",
             round(auc,digits=2),")",sep="")
 return(list(ninputs=ninputs,hpar=hpar,P=P,main=main))
}

# call to shinyApp (launches the app in browser):
shinyApp(ui=ui,server=server)
```

The Web demonstration is based on the **shiny** package discussed in Sect. 2.5. The interaction is obtained by selecting one solution from the Pareto front by using the **numericInput** component. To facilitate the visualization, the previously optimized Pareto front (stored in object **s1**) is first sorted according the number of input features. This is achieved by using the **sort.int** function, which sorts a vector and returns a list with the ordered indexes (at the **$ix** component). Each solution consists of a set of input features and hyperparameters that are shown in text output boxes (via the **shiny verbatimTextOutput** function). The **mainPainel** component also displays two plots (grouped horizontally via the **splitLayout** component): the Pareto front and the ROC curve computed using the wine quality test data. The Pareto plot highlights with the "X" symbol the currently selected model, while the ROC plot shows the target "high" quality class discrimination that is obtained for the holdout test data (**w[H$ts,]**).

Each time the user changes the selected Pareto point (by changing the **numericInput** component) the object **input$number** is updated and the reactive **modelInput** component is activated, calling the auxiliary **selectmodel** function. This produces changes in the Pareto plot ("X" point), ROC curve, and text outputs. The **selectmodel** function first decodes the selected genome (**x**). Then, it fits the associated XGBoost model (with the set of inputs and hyperparameter values that are defined by **x**) using all training data (via the **fit** function from **rminer**). Next, it computes the predictions for the test data (using the **predict rminer** function) and the respective AUC of the ROC curve (using the **mmetric rminer** function). Finally, the obtained results are grouped into a list that is

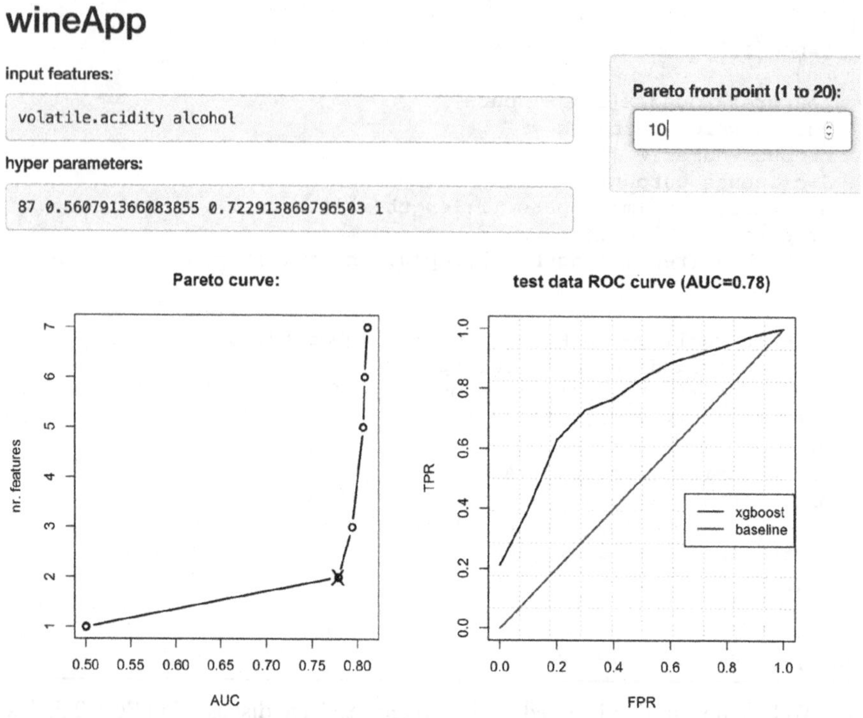

wineApp

input features:

volatile.acidity alcohol

hyper parameters:

87 0.560791366083855 0.722913869796503 1

Pareto front point (1 to 20):

10

Pareto curve:

test data ROC curve (AUC=0.78)

Fig. 7.7 Example of the wineApp execution

returned. In particular, the returned list is used by the component **output$roc**, allowing to plot the ROC curve via the **mgraph** function (from **rminer**). The Web demonstration can be launched in a browser by using:

```
> library(shiny)
> runApp("wineApp")
```

An example execution is shown in Fig. 7.7. In the example, a XGBoost model with 2 features was selected (**volatile.acidity** and **alcohol**). In the Pareto curve, the model is marked by the "X" symbol and it produced an AUC of 78% (when using the 3-fold validation over the training data). The associated AUC (shown in the ROC curve) for the test (unseen) data is also 78%, corresponding to a good discrimination (higher than 70%). Given that this model only uses two input variables, this is an interesting classifier that was found by the multi-objective search.

7.5 Command Summary

RCurl	package for network (HTTP/FTP/...) interface
TSP	package for traveling salesman problems
TSP()	creates a TSP instance (package **TSP**)
arima()	fit an ARIMA time series model
as.matrix()	convert to matrix
attr()	get or set an object attribute
auto.arima()	automatic identification and estimation of an ARIMA model (package **forecast**)
displacement()	displacement operator (chapter file **"perm.R"**)
dist()	computes a distance matrix between rows of a data matrix
eal()	permutation evolutionary algorithm with Lamarckian option (chapter file **"perm.R"**)
evaluateFitness()	evaluate a population (package **ecr**)
exchange()	exchange operator (chapter file **"perm.R"**)
fit()	fit a supervised data mining model (package **rminer**)
forecast	package for time series forecasting
forecast()	generic function for forecasting from a time series model (package **forecast**)
gArea()	compute the area of a polygon (package **rgeos**)
genPerm()	create a random population (package **ecr**)
generateOffspring()	breed an offspring (package **ecr**)
getURL()	download a URL (package **RCurl**)
holdout()	returns indexes for holdout data split with training and test sets (package **rminer**)
h5()	show a character using fifth level heading of the HTML format (package **shiny**)
initECRControl()	create a control object (package **ecr**)
insertion()	insertion operator (chapter file **"perm.R"**)
mining()	trains and tests a model under several runs and a given validation method (package **rminer**)
mgraph()	plots a data mining result graph (package **rminer**)
mmetric()	compute classification or regression error metrics (package **rminer**)
numericInput()	select a numeric value (package **shiny**)
ox()	order crossover (OX) operator (chapter file **"perm.R"**)
plot()	plot function for geometry objects (package **rgeos**)

`pmx()`	partially matched crossover (PMX) operator (chapter file `"perm.R"`)
`predict()`	predict function for fit objects (package **rminer**)
`reactive()`	sets a reactive code (package **shiny**)
`readWKT()`	read WKT format into a geometry object (package **rgeos**)
`registerECROperator()`	update a control object (package **ecr**)
`renderText()`	render a reactive text (package **shiny**)
`replaceMuCommaLambda()`	selection operation (package **ecr**)
`require()`	load a package
rgeos	package that interfaces to geometry engine—open source
rminer	package for a simpler use of classification and regression data mining methods
`scramble()`	scramble operator (chapter file `"perm.R"`)
`seq_len()`	generate a sequence (package **ecr**)
shiny	package for interactive web application
`solve_TSP()`	traveling salesman solver (package **TSP**)
`sort.int()`	sorts a vector, returning also the sorted indexes
`splitLayout()`	lays out elements horizontally (package **shiny**)
`strsplit()`	split elements of character vector
`verbatimTextOutput()`	render a verbatim text (package **shiny**)

7.6 Exercises

7.1 Encode the cycle crossover (`cx` function) for order representations, which performs a number of cycles between two parent solutions: P_1 and P_2 (Rocha et al., 2001). Cycle 1 starts with the first value in P_1 (v_1) and analyzes the value at same position in P_2 (v). Then, it searches for v in P_1 and analyzes the corresponding value at position P_2 (new v). This procedure continues until the new v is equal to v_1, ending the cycle. All P_1 and P_2 genes that were found in this cycle are marked. Cycle 2 starts with the first value in P_1 that is not marked and ends as described in cycle 1. The whole process proceeds with similar cycles until all genes have been marked. The genes marked in odd cycles are copied from P_1 to child 1 and from P_2 to child 2, while genes marked in even cycles are copied from P_1 to child 2 and from P_2 to child 1.

Show the offspring that result from applying the cycle crossover to the parents $P_1 = (1, 2, 3, 4, 5, 6, 7, 8, 9)$ and $P_2 = (9, 8, 1, 2, 3, 4, 5, 6, 7)$.

7.2 Using the same sunspots TSF example (from Sect. 7.3), optimize coefficients of the $ARIMA(2, 0, 1)$ model using a particle swarm optimization method and compare the MAE one-step ahead forecasts with the method returned by `auto.arima` function. As lower and upper bounds for the particle swarm optimization use the $[-1, 1]$ range for all coefficients of ARIMA except m, which should be searched around the sunspots average (within $\pm 10\%$ of the average value).

7.3 Adapt the wine classification code (**wine-quality.R** from Sect. 7.4), replacing the XGBoost model by a random forest (in one search) and a support vector machine with a Gaussian kernel (in another search). When using the **rminer** package, these classifiers are defined by setting the **mining** function argument: **model="randomForest"** and **model="ksvm"**. Consider two hyperparameters for the random forest: **ntree**—number of trees to grow (integer from 1 to 200); and **mtry**—number of inputs randomly sampled as candidates at each tree node split (integer from 1 to I, the number of inputs). Consider also two hyperparameters for the support vector machine: γ—the parameter value of the Gaussian kernel (real value from 2^{-6} to 2^{1}), in **rminer** this parameter is set by using **kpar=list(sigma=**γ**)**; and **C**—a penalty parameter (real value from 2^{-2} to 2^{5}). Finally, for each search, show the obtained Pareto front values (f_1, f_2) in a plot. Notes: for better results, the range for **ntree** should be higher (e.g., $\{1,2,\ldots,2000\}$) but a shorter range is used in this exercise to speed up the search; **rminer** package uses the default Gaussian **kernel="rbfdot"** but other kernels can be adopted (e.g., **"polydot"**—polynomial).

References

Alves NS, Mendes TS, de Mendonça MG, Spínola RO, Shull F, Seaman C (2016) Identification and management of technical debt: a systematic mapping study. Inf Softw Technol 70:100–121

Applegate D, Bixby R, Chvátal V, Cook W (2001) TSP cuts which do not conform to the template paradigm. In: Computational combinatorial optimization. Springer, pp 261–303

Applegate DL, Bixby RE, Chvatal V, Cook WJ (2011) The traveling salesman problem: a computational study. Princeton University Pres, Princeton

Bache K, Lichman M (2013) UCI machine learning repository. http://archive.ics.uci.edu/ml

Bäck T, Schwefel HP (1993) An overview of evolutionary algorithms for parameter optimization. Evol Comput 1(1):1–23

Baluja S (1994) Population-based incremental learning: a method for integrating genetic search based function optimization and competitive learning. Technical report, DTIC Document

Banzhaf W, Nordin P, Keller R, Francone F (1998) Genetic programming, an introduction. Morgan Kaufmann Publishers, Inc., San Francisco

Bélisle CJ (1992) Convergence theorems for a class of simulated annealing algorithms on Rd. J Appl Prob 29:885–895

Beume N, Naujoks B, Emmerich M (2007) SMS-EMOA: multiobjective selection based on dominated hypervolume. Eur J Oper Res 181(3):1653–1669

Beume N, Fonseca CM, López-Ibáñez M, Paquete L, Vahrenhold J (2009) On the complexity of computing the hypervolume indicator. IEEE Trans Evol Comput 13(5):1075–1082. https://doi.org/10.1109/TEVC.2009.2015575

Bossek J (2017) ecr 2.0: a modular framework for evolutionary computation in R. In: Proceedings of the genetic and evolutionary computation conference companion, pp 1187–1193

Boyd S, Vandenberghe L (2004) Convex optimization. Cambridge University Press, Cambridge

Brownlee J (2011) Clever algorithms: nature-inspired programming recipes. Jason Brownlee

Caflisch RE (1998) Monte Carlo and quasi-Monte Carlo methods. Acta Numer 1998:1–49

Cass S (2019) The top programming languages 2019. IEEE Spectrum. https://spectrum.ieee.org/computing/software/the-top-programming-languages-2019

Chen T, Guestrin C (2016) Xgboost: a scalable tree boosting system. In: Krishnapuram B, Shah M, Smola AJ, Aggarwal CC, Shen D, Rastogi R (eds) Proceedings of the 22nd ACM SIGKDD international conference on knowledge discovery and data mining, San Francisco, 13–17 Aug 2016. ACM, pp 785–794. https://doi.org/10.1145/2939672.2939785

Chen WN, Zhang J, Chung HS, Zhong WL, Wu WG, Shi YH (2010) A novel set-based particle swarm optimization method for discrete optimization problems. Evol Comput IEEE Trans 14(2):278–300

P. Cortez, *Modern Optimization with R*, Use R!,
https://doi.org/10.1007/978-3-030-72819-9

Ciupke K (2016) psoptim: Particle Swarm Optimization. https://CRAN.R-project.org/package=psoptim, R package version 1.0

Clerc M (2012) Standard particle swarm optimization. hal-00764996, version 1, http://hal.archives-ouvertes.fr/hal-00764996

Conceicao ELT (2016) DEoptimR: differential evolution optimization in pure R. https://CRAN.R-project.org/package=DEoptimR, R package version 1.0-8

Cortez P (2010) Data mining with neural networks and support vector machines using the R/rminer tool. In: Perner P (ed) Advances in data mining – applications and theoretical aspects, 10th industrial conference on data mining. LNAI 6171. Springer, Berlin, pp 572–583

Cortez P (2012) Data mining with multilayer perceptrons and support vector machines, chap 2. Springer, pp 9–25

Cortez P, Morais A (2007) A data mining approach to predict forest fires using meteorological data. In: Neves J et al (ed) New trends in artificial intelligence, 13th EPIA 2007 – Portuguese conference on artificial intelligence, APPIA, Guimarães, pp 512–523

Cortez P, Santos MF (2015) Recent advances on knowledge discovery and business intelligence. Expert Syst 32(3):433–434. https://doi.org/10.1111/exsy.12087

Cortez P, Cerdeira A, Almeida F, Matos T, Reis J (2009) Modeling wine preferences by data mining from physicochemical properties. Decis Support Syst 47(4):547–553

Cortez P, Pereira PJ, Mendes R (2020) Multi-step time series prediction intervals using neuroevolution. Neural Comput Appl 32(13):8939–8953. https://doi.org/10.1007/s00521-019-04387-3

Croes G (1958) A method for solving traveling-salesman problems. Oper Res 6(6):791–812

Deb K (2001) Multi-objective optimization using evolutionary algorithms. Wiley-Interscience series in systems and optimization. Wiley

Deb K, Jain H (2014) An evolutionary many-objective optimization algorithm using reference-point-based nondominated sorting approach, part I: solving problems with box constraints. IEEE Trans Evol Comput 18(4):577–601. https://doi.org/10.1109/TEVC.2013.2281535

Deb K, Sindhya K, Okabe T (2007) Self-adaptive simulated binary crossover for real-parameter optimization. In: Lipson H (ed) Genetic and evolutionary computation conference, GECCO 2007, Proceedings, London, 7–11 July 2007. ACM, pp 1187–1194. https://doi.org/10.1145/1276958.1277190

Dorigo M, Stützle T (2004) Ant colony optimization. MIT Press

Dorigo M, Maniezzo V, Colorni A (1996) Ant system: optimization by a colony of cooperating agents. IEEE Trans Syst Man Cybern Part B (Cybern) 26(1):29–41

Eberhart R, Kennedy J, Shi Y (2001) Swarm intelligence. Morgan Kaufmann

Eberhart RC, Shi Y (2011) Computational intelligence: concepts to implementations. Morgan Kaufmann

Fawcett T (2006) An introduction to ROC analysis. Pattern Recogn Lett 27:861–874

Fay C (2020) Companies, officials and NGO using R. https://github.com/ThinkR-open/companies-using-r

Fernandes G, Oliveira N, Cortez P, Mendes R (2020) A realistic scooter rebalancing system via metaheuristics. In: Coello CAC (ed) GECCO'20: genetic and evolutionary computation conference, companion Volume, Cancún, 8–12 July 2020. ACM, pp 265–266. https://doi.org/10.1145/3377929.3389905

Fernandes K, Vinagre P, Cortez P (2015) A proactive intelligent decision support system for predicting the popularity of online news. In: Pereira FC, Machado P, Costa E, Cardoso A (eds) Progress in artificial intelligence – 17th Portuguese conference on artificial intelligence, EPIA 2015, Coimbra, 8–11 Sept 2015. Proceedings. Lecture notes in computer science, vol 9273. Springer, pp 535–546. https://doi.org/10.1007/978-3-319-23485-4_53

Flasch O (2014) A friendly introduction to RGP. https://mran.microsoft.com/snapshot/2017-02-04/web/packages/rgp/vignettes/rgp_introduction.pdf

Freitas AA (2004) A critical review of multi-objective optimization in data mining: a position paper. ACM SIGKDD Explor Newslett 6(2):77–86

Gilli M, Maringer D, Schumann E (2019) Numerical methods and optimization in finance, 2nd edn. Academic Press

Glover F (1986) Future paths for integer programming and links to artificial intelligence. Comput Oper Res 13(5):533–549

Glover F (1990) Tabu search: a tutorial. Interfaces 20(4):74–94

Glover F, Laguna M (1998) Tabu search. Springer

Goldberg DE, Deb K (1990) A comparative analysis of selection schemes used in genetic algorithms. In: Rawlins GJE (ed) Proceedings of the first workshop on foundations of genetic algorithms, Bloomington Campus, 15–18 July 1990. Morgan Kaufmann, pp 69–93. https://doi.org/10.1016/b978-0-08-050684-5.50008-2

Gonçalves S, Cortez P, Moro S (2020) A deep learning classifier for sentence classification in biomedical and computer science abstracts. Neural Comput Appl 32(11):6793–6807. https://doi.org/10.1007/s00521-019-04334-2

Gonzalez-Fernandez Y, Soto M (2014) copulaedas: an R package for estimation of distribution algorithms based on copulas. J Stat Softw 58(9):1–34. http://www.jstatsoft.org/v58/i09/

Guyon I, Elisseeff A (2003) An introduction to variable and feature selection. J Mach Learn Res 3:1157–1182

Hashimoto R, Ishibuchi H, Masuyama N, Nojima Y (2018) Analysis of evolutionary multi-tasking as an island model. In: Aguirre HE, Takadama K (eds) Proceedings of the genetic and evolutionary computation conference companion, GECCO 2018, Kyoto, 15–19 July 2018. ACM, pp 1894–1897. https://doi.org/10.1145/3205651.3208228

Holland J (1975) Adaptation in natural and artificial systems. PhD thesis, University of Michigan, Ann Arbor

Hsu CW, Chang CC, Lin CJ (2003) A practical guide to support vector classification

Huang CM, Lee YJ, Lin DK, Huang SY (2007) Model selection for support vector machines via uniform design. Comput Stat Data Anal 52(1):335–346

Huband S, Hingston P, Barone L, While L (2006) A review of multiobjective test problems and a scalable test problem toolkit. Evol Comput IEEE Trans 10(5):477–506

Hutter F, Kotthoff L, Vanschoren J (eds) (2019) Automated machine learning – methods, systems, challenges. The Springer series on challenges in machine learning. Springer. https://doi.org/10.1007/978-3-030-05318-5

Igel C (2014) No free lunch theorems: limitations and perspectives of metaheuristics. In: Borenstein Y, Moraglio A (eds) Theory and principled methods for the design of metaheuristics. Natural computing series. Springer, pp 1–23. https://doi.org/10.1007/978-3-642-33206-7_1

Ihaka R, Gentleman R (1996) R: a language for data analysis and graphics. J Comput Graph Stat 5(3):299–314

Irawan D, Naujoks B (2019) Comparison of reference- and hypervolume-based MOEA on solving many-objective optimization problems. In: Deb K, Goodman ED, Coello CAC, Klamroth K, Miettinen K, Mostaghim S, Reed P (eds) Evolutionary multi-criterion optimization – 10th international conference, EMO 2019, East Lansing, 10–13 Mar 2019, Proceedings. Lecture notes in computer science, vol 11411. Springer, pp 266–277, https://doi.org/10.1007/978-3-030-12598-1_22

Joe H (1997) Multivariate models and dependence concepts, vol 73. CRC Press

Kaboudan MA (2003) Forecasting with computer-evolved model specifications: a genetic programming application. Comput Oper Res 30(11):1661–1681

Kennedy J, Eberhart R (1995) Particle swarm optimization. In: ICNN'95 – IEEE international conference on neural networks proceedings. IEEE Computer Society, Perth, pp 1942–1948

Kirkpatrick S, Gelatt CD, Vecchi MP (1983) Optimization by simulated annealing. Science 220:671–680

Koch R (2015) From business intelligence to predictive analytics. Strategic Financ 96(7):56

Kohavi R (1995) A study of cross-validation and bootstrap for accuracy estimation and model selection. In: Proceedings of the international joint conference on artificial intelligence (IJCAI), Montreal, vol 2. Morgan Kaufmann

Konak A, Coit DW, Smith AE (2006) Multi-objective optimization using genetic algorithms: a tutorial. Reliab Eng Syst Saf 91(9):992–1007

Larrañaga P, Lozano JA (2002) Estimation of distribution algorithms: a new tool for evolutionary computation, vol 2. Springer, New York

Liagkouras K, Metaxiotis K (2013) An elitist polynomial mutation operator for improved performance of moeas in computer networks. In: 22nd international conference on computer communication and networks, ICCCN 2013, Nassau, 30 July–2 Aug 2013. IEEE, pp 1–5. https://doi.org/10.1109/ICCCN.2013.6614105

López-Ibáñez M, Dubois-Lacoste J, Cáceres LP, Birattari M, Stützle T (2016) The irace package: iterated racing for automatic algorithm configuration. Oper Res Perspect 3:43–58

Lucasius CB, Kateman G (1993) Understanding and using genetic algorithms part 1. Concepts, properties and context. Chemom Intell Lab Syst 19(1):1–33

Luke S (2015) Essentials of metaheuristics. Lulu.com, online version 2.2 at http://cs.gmu.edu/~sean/book/metaheuristics

Makridakis S, Weelwright S, Hyndman R (1998) Forecasting: methods and applications, 3rd edn. Wiley, New York

Mendes R (2004) Population topologies and their influence in particle swarm performance. PhD thesis, Universidade do Minho

Mendes R, Cortez P, Rocha M, Neves J (2002) Particle swarms for feedforward neural network training. In: Proceedings of the 2002 international joint conference on neural networks (IJCNN 2002), Honolulu. IEEE Computer Society, pp 1895–1899

Mendes R, Kennedy J, Neves J (2004) The fully informed particle swarm: simpler, maybe better. IEEE Trans Evol Comput 8(3):204–210. https://doi.org/10.1109/TEVC.2004.826074

Michalewicz Z (1996) Genetic algorithms + data structures = evolution programs. Springer, New York

Michalewicz Z (2008) Adaptive Business Intelligence, Computer Science Course 7005 Handouts

Michalewicz Z, Fogel D (2004) How to solve it: modern heuristics. Springer, New York

Michalewicz Z, Schmidt M, Michalewicz M, Chiriac C (2006) Adaptive business intelligence. Springer, New York

Michalewicz Z, Schmidt M, Michalewicz M, Chiriac C (2007) Adaptive business intelligence: three case studies. Evol Comput Dyn Uncertain Environ 51:179–196

Muenchen RA (2019) The popularity of data analysis software. http://r4stats.com/articles/popularity/

Mühlenbein H (1997) The equation for response to selection and its use for prediction. Evol Comput 5(3):303–346

Mullen K, Ardia D, Gil D, Windover D, Cline J (2011) Deoptim: an R package for global optimization by differential evolution. J Stat Softw 40(6):1–26

Nagata Y, Kobayashi S (2013) A powerful genetic algorithm using edge assembly crossover for the traveling salesman problem. INFORMS J Comput 25(2):346–363. https://doi.org/10.1287/ijoc.1120.0506

Noorian F, de Silva AM, Leong PHW (2016) gramEvol: grammatical evolution in R. J Stat Softw 71(1):1–26. https://doi.org/10.18637/jss.v071.i01

Paradis E (2002) R for beginners. Montpellier (F): University of Montpellier. https://cran.r-project.org/doc/contrib/Paradis-rdebuts_en.pdf

Parente M, Cortez P, Correia AG (2015) An evolutionary multi-objective optimization system for earthworks. Expert Syst Appl 42(19):6674–6685. https://doi.org/10.1016/j.eswa.2015.04.051

Pereira PJ, Pinto P, Mendes R, Cortez P, Moreau A (2019) Using neuroevolution for predicting mobile marketing conversion. In: Progress in artificial intelligence, 19th EPIA conference on artificial intelligence, EPIA 2019, Vila Real, 3–6 Sept 2019, Proceedings, Part II. Lecture notes in computer science, vol 11805. Springer, pp 373–384. https://doi.org/10.1007/978-3-030-30244-3_31

Price KV, Storn RM, Lampinen JA (2005) Differential evolution a practical approach to global optimization. Springer, New York

Quantargo (2020) R is everywhere. https://www.r-bloggers.com/r-is-everywhere/

R Core Team (2020) R: a language and environment for statistical computing. R Foundation for Statistical Computing, Vienna. https://www.R-project.org/

Reinelt G (1994) The traveling salesman: computational solutions for TSP applications. Springer

Robert C, Casella G (2009) Introducing Monte Carlo methods with R. Springer, New York

Rocha M, Cortez P, Neves J (2000) The relationship between learning and evolution in static and in dynamic environments. In: Fyfe C (ed) Proceedings of the 2nd ICSC symposium on engineering of intelligent systems (EIS'2000). ICSC Academic Press, pp 377–383

Rocha M, Mendes R, Cortez P, Neves J (2001) Sitting guest at a wedding party: experiments on genetic and evolutionary constrained optimization. In: Proceedings of the 2001 congress on evolutionary computation (CEC2001), vol 1. IEEE Computer Society, Seoul, pp 671–678

Rocha M, Cortez P, Neves J (2007) Evolution of neural networks for classification and regression. Neurocomputing 70:2809–2816

Rocha M, Sousa P, Cortez P, Rio M (2011) Quality of service constrained routing optimization using evolutionary computation. Appl Soft Comput 11(1):356–364

Rudolph G, Schütze O, Grimme C, Trautmann H (2014) An aspiration set EMOA based on averaged hausdorff distances. In: Pardalos PM, Resende MGC, Vogiatzis C, Walteros JL (eds) Learning and intelligent optimization – 8th international conference, Lion 8, Gainesville, 16–21 Feb 2014. Revised selected papers. Lecture notes in computer science, vol 8426. Springer, pp 153–156. https://doi.org/10.1007/978-3-319-09584-4_15

Ryan C, Collins JJ, O'Neill M (1998) Grammatical evolution: evolving programs for an arbitrary language. In: Banzhaf W, Poli R, Schoenauer M, Fogarty TC (eds) Genetic programming, first European workshop, EuroGP'98, Paris, 14–15 Apr 1998, Proceedings. Lecture notes in computer science, vol 1391. Springer, pp 83–96. https://doi.org/10.1007/BFb0055930

Satman MH (2013) Machine coded genetic algorithms for real parameter optimization problems. Gazi Univ J Sci 26(1):85–95

Schrijver A (1998) Theory of linear and integer programming. Wiley, Chichester

Scrucca L (2017) On some extensions to GA package: hybrid optimisation, parallelisation and islands evolution. R J 9(1):187–206. https://journal.r-project.org/archive/2017/RJ-2017-008

Socha K, Dorigo M (2008) Ant colony optimization for continuous domains. Eur J Oper Res 185(3):1155–1173

Sörensen K (2015) Metaheuristics – the metaphor exposed. ITOR 22(1):3–18. https://doi.org/10.1111/itor.12001

Stepnicka M, Cortez P, Donate JP, Stepnicková L (2013) Forecasting seasonal time series with computational intelligence: on recent methods and the potential of their combinations. Expert Syst Appl 40(6):1981–1992

Storn R, Price K (1997) Differential evolution–a simple and efficient heuristic for global optimization over continuous spaces. J Global Optim 11(4):341–359

Tang K, Li X, Suganthan P, Yang Z, Weise T (2009) Benchmark functions for the cec'2010 special session and competition on large-scale global optimization. Technical report, University of Science and Technology of China

Trivedi V, Varshney P, Ramteke M (2020) A simplified multi-objective particle swarm optimization algorithm. Swarm Intell 14(2):83–116. https://doi.org/10.1007/s11721-019-00170-1

Tsallis C, Stariolo DA (1996) Generalized simulated annealing. Phys A Stat Mech Appl 233(1–2):395–406

Turban E, Sharda R, Aronson J, King D (2010) Business intelligence, A managerial approach, 2nd edn. Prentice-Hall

Vance A (2009) R you ready for R? http://bits.blogs.nytimes.com/2009/01/08/r-you-ready-for-r/

Venables W, Smith D, R Core Team (2013) An introduction to R. http://cran.r-project.org/doc/manuals/R-intro.pdf

Wolpert DH, Macready WG (1997) No free lunch theorems for optimization. Evol Comput IEEE Trans 1(1):67–82

Xiang Y, Gubian S, Suomela B, Hoeng J (2013) Generalized simulated annealing for global optimization: the GenSA package. R J 5(1):13

Xu Q, Xu Z, Ma T (2020) A survey of multiobjective evolutionary algorithms based on decomposition: variants, challenges and future directions. IEEE Access 8:41588–41614. https://doi.org/10.1109/ACCESS.2020.2973670

Yang XS (2014) Nature-inspired optimization algorithms. Elsevier, Waltham

Zuur A, Ieno E, Meesters E (2009) A beginner's guide to R. Springer, New York

Solutions

Exercises of Chapter 2

2.1

```
v=rep(0,10)    # same as: v=vector(length=10);v[]=0
v[c(3,7,9)]=1 # update values
print(v)       # show v
```

2.3

```
v=seq(2,50,by=2) # one way
print(v)
v=(1:25)*2 # other way
print(v)
```

2.3

```
m=matrix(nrow=3,ncol=4)
m[1,]=1:4
m[2,]=sqrt(m[1,])
m[3,]=sqrt(m[2,])
m[,4]=m[,3]^2 # m[,3]*m[,3]
print(round(m,digits=2))
cat("sums of rows:",round(apply(m,1,sum),digits=2),"\n")
cat("sums of columns:",round(apply(m,2,sum),digits=2),"\n")
```

2.4

```
# 1 - use of for ... if
counteven1=function(x)
{ r=0
  for(i in 1:length(x))
     { if(x[i]%%2==0) r=r+1 }
  return(r)
}
```


P. Cortez, *Modern Optimization with R*, Use R!,
https://doi.org/10.1007/978-3-030-72819-9

```
# 2 - use of sapply
# auxiliary function
ifeven=function(x) # x is a number
{ if(x%%2) return(TRUE) else return(FALSE)}

counteven2=function(x)
{ return(sum(sapply(x,ifeven))) }

# 3 - use of direct condition (easiest way)
counteven3=function(x)
{ return(sum(x%%2==0)) }

x=1:10
cat("counteven1:",counteven1(x),"\n")
cat("counteven2:",counteven2(x),"\n")
cat("counteven3:",counteven3(x),"\n")
```

2.5

```
DIR="" # change to other directory if needed
pdf(paste(DIR,"maxsin.pdf",sep=""),width=5,height=5) # create PDF
D=8 # number of binary digits, the dimension
x=0:(2^D-1);y=sin(pi*x/2^D)
plot(x,y,type="l",ylab="evaluation function",
     xlab="search space",lwd=2)
pmax=c(x[which.max(y)],max(y)) # set the maximum point
points(pmax[1],pmax[2],pch=19,lwd=2) # plot the maximum
legend("topright","optimum",pch=19,lwd=2) # add a legend
dev.off() # close the graphical device
```

2.6

```
# 1
# install.packages("RCurl") # if needed, install the package
library(RCurl)
# 2
fires=getURL("http://archive.ics.uci.edu/ml/
    machine-learning-databases/forest-fires/forestfires.csv")
write(fires,file="forestfires.csv") # write to working directory
# 3, read file:
fires=read.table("forestfires.csv",header=TRUE,sep=",")
# 4
aug=fires$temp[fires$month=="aug"]
cat("mean temperature in Aug.:",mean(aug),"\n")
# 5
feb=fires$temp[fires$month=="feb"]
jul=fires$temp[fires$month=="jul"]
sfeb=sample(feb,10)
sjul=sample(jul,10)
saug=sample(aug,10)
p1=t.test(saug,sfeb)$p.value
p2=t.test(saug,sjul)$p.value
p3=t.test(sjul,sfeb)$p.value
```

```
cat("p-values (Aug-Feb,Aug-Jul,Jul-Feb):",
    round(c(p1,p2,p3),digits=2),"\n")
# 6
aug100=fires[fires$month=="aug"&fires$area>100,]
print(aug100)
# 7
write.table(aug100,"aug100.csv",sep=",",row.names=FALSE)
```

Exercises of Chapter 3

3.1

```
source("blind.R") # load the blind search methods

binint=function(x,D)
{ x=rev(intToBits(x)[1:D]) # get D bits
  # remove extra 0s from raw type:
  as.numeric(unlist(strsplit(as.character(x),""))[(1:D)*2])
}
intbin=function(x) sum(2^(which(rev(x==1))-1))
maxsin=function(x,Dim) sin(pi*(intbin(x))/(2^Dim))

D=16 # number of dimensions

# blind search:
PTM=proc.time() # start clock
x=0:(2^D-1) # integer search space
search=t(sapply(x,binint,D=D))
S=fsearch(search,maxsin,"max",D) # full search
sec=(proc.time()-PTM)[3] # get seconds elapsed
cat("fsearch s:",S$sol,"f:",S$eval,"time:",sec,"s\n")

# adapted grid search:
N=1000
PTM=proc.time() # start clock
x=seq(0,2^D-1,length.out=N)
search=t(sapply(x,binint,D=D))
S=fsearch(search,maxsin,"max",D) # grid
sec=(proc.time()-PTM)[3] # get seconds elapsed
cat("gsearch s:",S$sol,"f:",S$eval,"time:",sec,"s\n")

# adapted monte carlo search:
PTM=proc.time() # start clock
x=sample(0:2^D-1,N)
search=t(sapply(x,binint,D=D))
S=fsearch(search,maxsin,"max",D) # grid
sec=(proc.time()-PTM)[3] # get seconds elapsed
cat("mcsearch s:",S$sol,"f:",S$eval,"time:",sec,"s\n")
```

3.2

```
source("blind.R") # load the blind search methods
source("grid.R") # load the grid search methods
source("functions.R") # load the profit function

D=5 # number of dimensions
# grid search code:
S1=gsearch(profit,rep(350,D),rep(450,D),rep(11,D),"max")
cat("gsearch s:",round(S1$sol),"f:",S1$eval,"\n")

# dfsearch code:
domain=vector("list",D)
for(i in 1:D) domain[[i]]=seq(350,450,by=11)
S2=dfsearch(domain=domain,fn=profit,type="max")
cat("dfsearch s:",round(S2$sol),"f:",S2$eval,"\n")
```

3.3

```
source("blind.R") # load the blind search methods
source("montecarlo.R") # load the monte carlo method

rastrigin=function(x) 10*length(x)+sum(x^2-10*cos(2*pi*x))

# experiment setup parameters:
D=30
Runs=30
N=10^c(2,3,4) # number of samples

# perform all monte carlo searches:
S=matrix(nrow=Runs,ncol=length(N))
for(j in 1:length(N)) # cycle all number of samples
for(i in 1:Runs) # cycle all runs
   S[i,j]=mcsearch(rastrigin,rep(-5.2,D),rep(5.2,D),
                   N[j],"min")$eval
# compare average results:
p21=t.test(S[,2],S[,1])$p.value
p31=t.test(S[,3],S[,2])$p.value
cat("N=",N,"\n")
cat("average f:",apply(S,2,mean),"\n")
cat("p-value (N=",N[2],"vs N=",N[1],")=",
    round(p21,digits=2),"\n")
cat("p-value (N=",N[3],"vs N=",N[2],")=",
    round(p31,digits=2),"\n")
boxplot(S[,1],S[,2],S[,3],names=paste("N=",N,sep=""))
```

Exercises of Chapter 4

4.1

```
source("hill.R")

# steepest ascent and stochastic hill climbing:
#    par - initial solution
#    fn - evaluation function
#    change - function to generate the next candidate
#    lower - vector with lowest values for each dimension
#    upper - vector with highest values for each dimension
#    control - list with stopping and monitoring method:
#       $N - number of change searches (steepest ascent)
#       $P - Probability (in [0,1]) for accepting solutions
#       $maxit - maximum number of iterations
#       $REPORT - frequency of monitoring information
#       $digits - (optional) round digits for reporting
#    type - "min" or "max"
#    ... - extra parameters for fn
s2_hclimbing=function(par,fn,change,lower,upper,control,
                  type="min",...)
{ fpar=fn(par,...)
  b=list(par=par,fpar=fpar) # initial best
  for(i in 1:control$maxit)
     {
      # first change
      par1=change(par,lower,upper)
      fpar1=fn(par1,...)
      if(control$N>1) # steepest ascent cycle
      { for(j in 1:(control$N-1))
         { # random search for better par1 solutions:
          par2=change(par,lower,upper)
          fpar2=fn(par2,...)
          b1=best(par1,fpar1,par2,fpar2,type)
          par1=b1$par;fpar1=b1$fpar # update change
         }
      }

      if(control$REPORT>0 &&(i==1||i%%control$REPORT==0))
         report_iter(i,par,fpar,par1,fpar1,control)

      b=best(b$par,b$fpar,par1,fpar1,type) # memorize best

      x=runif(1) # random between [0,1]
      if(x<control$P) # accept new solution
       { par=par1; fpar=fpar1 }
      else # select best between par and par1
       {
        b1=best(par,fpar,par1,fpar1,type)
        par=b1$par;fpar=b1$fpar # update par
       }
     }
```

```
  par=b$par;fpar=b$fpar # set par to best
  if(control$REPORT>=1)
     report_iter("best:",par,fpar,control=control)
  return(list(sol=par,eval=fpar))
}
```

4.2

```
source("hill.R") # load the hill climbing methods
library(tabuSearch) # load tabuSearch

intbin=function(x) sum(2^(which(rev(x==1))-1))
maxsin=function(x) sin(pi*(intbin(x))/(2^D))
D=16 # number of dimensions
s=rep(0,D) # initial search point

# hill climbing:
maxit=20
C=list(maxit=maxit,REPORT=0) # maximum of 10 iterations
ichange=function(par,lower,upper) # integer change
{hchange(par,lower,upper,rnorm,mean=0,sd=1) }
b=hclimbing(s,maxsin,change=ichange,lower=rep(0,D),upper=rep(1,D)
    ,
          control=C,type="max")
cat("hill b:",b$sol,"f:",b$eval,"\n")

# simulated annealing:
eval=function(x) -maxsin(x)
ichange2=function(par) # integer change
{D=length(par);hchange(par,lower=rep(0,D),upper=rep(1,D),rnorm,
    mean=0,sd=1)}
C=list(maxit=maxit)
b=optim(s,eval,method="SANN",gr=ichange2,control=C)
cat("sann b:",b$par,"f:",abs(b$value),"\n")

# tabu search:
b=tabuSearch(size=D,iters=maxit,objFunc=maxsin,config=s,neigh=4,
    listSize=8)
ib=which.max(b$eUtilityKeep) # best index
cat("tabu b:",b$configKeep[ib,],"f:",b$eUtilityKeep[ib],"\n")
```

4.3

```
library(tabuSearch) # get tabuSearch

rastrigin=function(x) f=10*length(x)+sum(x^2-10*cos(2*pi*x))
intbin=function(x) # convert binary to integer
{ sum(2^(which(rev(x==1))-1)) } # explained in Chapter 3
breal=function(x) # convert binary to D real values
{ # note: D and bits need to be set outside this function
  s=vector(length=D)
  for(i in 1:D) # convert x into s:
  { ini=(i-1)*bits+1;end=ini+bits-1
    n=intbin(x[ini:end])
```

```
    s[i]=lower+n*drange/2^bits
  }
  return(s)
}
# note: tabuSearch does not work well with negative evaluations
# to solve this drawback, a MAXIMUM constant is defined
MAXIMUM=10000
brastrigin=function(x) MAXIMUM-rastrigin(breal(x)) # max. goal

D=8;MAXIT=500
bits=8 # per dimension
size=D*bits
lower=-5.2;upper=5.2;drange=upper-lower
s=sample(0:1,size=size,replace=TRUE)
b=tabuSearch(size=size,iters=MAXIT,objFunc=brastrigin,config=s,
    neigh=bits,listSize=bits,nRestarts=1)
ib=which.max(b$eUtilityKeep) # best index
cat("b:",b$configKeep[ib,],"f:",MAXIMUM-b$eUtilityKeep[ib],"\n")
```

4.4

```
library(irace)
library(tabuSearch) # load tabuSearch package
source("functions.R") # load the profit function

# tabu search for bag prices:
D=5 # dimension (number of prices)
MaxPrice=1000
Dim=ceiling(log(MaxPrice,2)) # size of each price (=10)
size=D*Dim # total number of bits (=50)
s0=sample(0:1,size,replace=TRUE) # initial search

intbin=function(x) # convert binary to integer
{ sum(2^(which(rev(x==1))-1)) } # explained in Chapter 3

bintbin=function(x) # convert binary to D prices
{ # note: D and Dim need to be set outside this function
  s=vector(length=D)
  for(i in 1:D) # convert x into s:
  { ini=(i-1)*Dim+1;end=ini+Dim-1
    s[i]=intbin(x[ini:end])
  }
  return(s)
}

bprofit=function(x) # profit for binary x
{ s=bintbin(x)
  if(sum(s>MaxPrice)>0) f=-Inf # death penalty
  else f=profit(s)
  return(f)
}

# irace parameters:
#  neigh from 1 to size
```

```
#  listSize from 2 (minimum) to twice the default (2*9=18)
parameters.txt='
neigh "" i (1, 50)
listSize "" i (2, 18)
'
parameters=readParameters(text=parameters.txt)

cat("initial:",bintbin(s0),"f:",bprofit(s0),"\n")

# evaluate each irace configuration:
runner=function(experiment,scenario)
{
 C=experiment$configuration # get current irace configuration
 s=tabuSearch(size,iters=100,objFunc=bprofit,config=s0,nRestarts
    =1, # fixed part
             neigh=C$neigh,listSize=C$listSize) # irace part
 b=which.max(s$eUtilityKeep) # best index
 # since tabuSearch performs maximization, return -profit:
 return(list(cost=-1*s$eUtilityKeep[b]))
}

scenario=list(targetRunner=runner,
              instances=1, # not used but needs to be defined
              maxExperiments=100, # 100 calls to targetRunner
              logFile = "") # do not create log file

ir=irace(scenario=scenario,parameters=parameters)
configurations.print(ir)
```

Exercises of Chapter 5

5.1

```
library(genalg) # load genalg
library(ecr)    # load ecr
library(GA)     # load GA
library(NMOF)   # load GAopt

# sum of bits:
maxsumbin=function(x) sum(x)
minsumbin=function(x) (length(x)-sum(x))

D=20 # number of dimensions

# GA parameter setup:
Pop=20 # population size
Gen=100 # maximum number of generations
Eli=1L # elitism of one individual

# auxiliary function:
showres=function(x)
```

```
{ cat("x:",x,"f:",maxsumbin(x),"\n") }

# rbga.bin (minimization):
cat("rbga.bin:\n")
ga1=rbga.bin(size=D,popSize=Pop,iters=Gen,
             evalFunc=minsumbin,elitism=Eli)
b=which.min(ga1$evaluations) # best individual
# show best solution and evaluation value:
showres(ga1$population[b,])

# ecr (maximization):
cat("ecr:\n")
ga2=ecr(maxsumbin,minimize=FALSE,n.objectives=1,
        n.dim=D,n.bits=D,representation="binary",
        mu=Pop, # population size
        lambda=(Pop-Eli), # new solutions in each generation
        n.elite=Eli,
        terminators = list(stopOnIters(Gen)))
showres(ga2$best.x[[1]])

# ga (maximization):
cat("ga:\n")
ga3=ga(type="binary",maxsumbin,nBits=D,
       popSize=Pop,elitism=Eli,maxiter=Gen,monitor=FALSE)
showres(ga3@solution)

# GAopt (minimization):
cat("GAopt:\n")
algo=list(nB=D,nP=Pop,nG=Gen,
          printDetail=FALSE,printBar=FALSE)
ga4=GAopt(minsumbin,algo=algo)
showres(as.numeric(ga4$xbest))
```

5.2

```
library(pso)
library(copulaedas)
source("blind.R") # get fsearch
source("montecarlo.R") # get mcsearch

# evaluation function: ------------------------------------
eggholder=function(x) # length of x is 2
{ x=ifelse(x<lower[1],lower[1],x) # (only due to EDA):
  x=ifelse(x>upper[1],upper[1],x) # bound if needed
  f=(-(x[2]+47)*sin(sqrt(abs(x[2]+x[1]/2+47)))
     -x[1]*sin(sqrt(abs(x[1]-(x[2]+47))))
    )
  # global assignment code: <<-
  EV<<-EV+1 # increase evaluations
  if(f<BEST) BEST<<-f # minimum value
  if(EV<=MAXFN) F[EV]<<-BEST
  return(f)
}
```

```
# auxiliary functions: -----------------------------------
crun2=function(method,f,lower,upper,LP,maxit,MAXFN) # run a
    method
{ if(method=="MC")
   {
    s=runif(D,lower[1],upper[1]) # initial search point
    mcsearch(fn=eggholder,lower=lower,upper=upper,N=MAXFN)
   }
  else if(method=="PSO")
     { C=list(maxit=maxit,s=LP,type="SPSO2011")
       psoptim(rep(NA,length(lower)),fn=f,
               lower=lower,upper=upper,control=C)
     }
  else if(method=="EDA")
     { setMethod("edaTerminate","EDA",edaTerminateMaxGen)
       DVEDA=VEDA(vine="DVine",indepTestSigLevel=0.01,
           copulas = c("normal"),margin = "norm")
       DVEDA@name="DVEDA"
       edaRun(DVEDA,f,lower,upper)
     }
}

successes=function(x,LIM,type="min") # number of successes
{ if(type=="min") return(sum(x<LIM)) else return(sum(x>LIM)) }

ctest2=function(Methods,f,lower,upper,type="min",Runs, # test
               D,MAXFN,maxit,LP,pdf,main,LIM) # all methods:
{ RES=vector("list",length(Methods)) # all results
  VAL=matrix(nrow=Runs,ncol=length(Methods)) # best values
  for(m in 1:length(Methods)) # initialize RES object
   RES[[m]]=matrix(nrow=MAXFN,ncol=Runs)

  for(R in 1:Runs) # cycle all runs
    for(m in 1:length(Methods))
      { EV<<-0; F<<-rep(NA,MAXFN) # reset EV and F
        if(type=="min") BEST<<-Inf else BEST<<- -Inf # reset BEST
        suppressWarnings(crun2(Methods[m],f,lower,upper,LP,maxit,
            MAXFN))
        RES[[m]][,R]=F # store all best values
        VAL[R,m]=F[MAXFN] # store best value at MAXFN
      }
  # compute average F result per method:
  AV=matrix(nrow=MAXFN,ncol=length(Methods))
  for(m in 1:length(Methods))
    for(i in 1:MAXFN)
      AV[i,m]=mean(RES[[m]][i,])
  # show results:
  cat(main,"\n",Methods,"\n")
  cat(round(apply(VAL,2,mean),digits=0)," (average best)\n")
  cat(round(100*apply(VAL,2,successes,LIM,type)/Runs,
          digits=0)," (%successes)\n")

  # create pdf file:
  pdf(paste(pdf,".pdf",sep=""),width=5,height=5,paper="special")
```

```
  par(mar=c(4.0,4.0,1.8,0.6)) # reduce default plot margin
  MIN=min(AV);MAX=max(AV)
  # use a grid to improve clarity:
  g1=seq(1,MAXFN,length.out=500) # grid for lines
  plot(g1,AV[g1,1],ylim=c(MIN,MAX),type="l",lwd=2,main=main,
       ylab="average best",xlab="number of evaluations")
  for(i in 2:length(Methods)) lines(g1,AV[g1,i],lwd=2,lty=i)
  if(type=="min") position="topright" else position="bottomright"
  legend(position,legend=Methods,lwd=2,lty=1:length(Methods))
  dev.off() # close the PDF device
}

# define EV, BEST and F:
MAXFN=1000
EV=0;BEST=Inf;F=rep(NA,MAXFN)
# define method labels:
Methods=c("MC","PSO","EDA")
# eggholder comparison: ----------------------------------
Runs=10; D=2; LP=20; maxit=50
lower=rep(-512,D);upper=rep(512,D)
ctest2(Methods,eggholder,lower,upper,"min",Runs,D,MAXFN,maxit,LP,
      "comp-eggholder","eggholder (D=2)",-950)
```

5.3

```
source("functions.R") # bag prices functions
library(copulaedas) # EDA

# auxiliary functions: ------------------------------------

# returns TRUE if prices are sorted in descending order
prices_ord=function(x)
{ d=diff(x) # d lagged differences x(i+1)-x(i)
  if(sum(d>=0)) return (FALSE) else return (TRUE)
}
ord_prices=function(x)
{ x=sort(x,decreasing=TRUE) # sort x
  # x is sorted but there can be ties:
  k=2           # remove ties by removing $1
  while(!prices_ord(x)) # at each iteration
    { if(x[k]==x[k-1]) x[k]=x[k]-1
      k=k+1
    }
  return(x)
}

# evaluation function: ------------------------------------
cprofit3=function(x) # bag prices with death penalty
{ x=round(x,digits=0) # convert x into integer
  x=ifelse(x<1,1,x)         # assure that x is within
  x=ifelse(x>1000,1000,x)  # the [1,1000] bounds
  if(!prices_ord(x)) res=Inf # if needed, death penalty!!!
  else
    {
```

```
      s=sales(x);c=cost(s);profit=sum(s*x-c)
      # if needed, store best value
      if(profit>BEST) { BEST<<-profit; B<<-x}
      res=-profit # minimization task!
     }
   EV<<-EV+1 # increase evaluations
   if(EV<=MAXFN) F[EV]<<-BEST
   return(res)
}
# example of a very simple and fast repair of a solution:
# sort the solution values!
localRepair2=function(eda, gen, pop, popEval, f, lower, upper)
{
 for(i in 1:nrow(pop))
 { x=pop[i,]
   x=round(x,digits=0) # convert x into integer
   x=ifelse(x<lower[1],lower[1],x) # assure x within
   x=ifelse(x>upper[1],upper[1],x) # bounds
   if(!prices_ord(x)) x=ord_prices(x) # order x
   pop[i,]=x;popEval[i]=f(x) # replace x in population
 }
 return(list(pop=pop,popEval=popEval))
}

# experiment: ----------------------------------------------
MAXFN=5000
Runs=50; D=5; LP=50; maxit=100
lower=rep(1,D);upper=rep(1000,D)
Methods=c("Death","Repair")
setMethod("edaTerminate","EDA",edaTerminateMaxGen)
UMDA=CEDA(copula="indep",margin="norm"); UMDA@name="UMDA"

RES=vector("list",length(Methods)) # all results
VAL=matrix(nrow=Runs,ncol=length(Methods)) # best values
for(m in 1:length(Methods)) # initialize RES object
   RES[[m]]=matrix(nrow=MAXFN,ncol=Runs)
for(R in 1:Runs) # cycle all runs
  {
    B=NA;EV=0; F=rep(NA,MAXFN); BEST= -Inf # reset vars.
    setMethod("edaOptimize","EDA",edaOptimizeDisabled)
    setMethod("edaTerminate","EDA",edaTerminateMaxGen)
    suppressWarnings(edaRun(UMDA,cprofit3,lower,upper))
    RES[[1]][,R]=F # store all best values
    VAL[R,1]=F[MAXFN] # store best value at MAXFN

    B=NA;EV=0; F=rep(NA,MAXFN); BEST= -Inf # reset vars.
    # set local repair search method:
    setMethod("edaOptimize","EDA",localRepair2)
    # set additional termination criterion:
    setMethod("edaTerminate","EDA",
              edaTerminateCombined(edaTerminateMaxGen,
                  edaTerminateEvalStdDev))
    # this edaRun might produces warnings or errors:
```

```
    suppressWarnings(try(edaRun(UMDA,cprofit3,lower,upper),silent
        =TRUE))
    if(EV<MAXFN) # if stopped due to EvalStdDev
       F[(EV+1):MAXFN]=rep(F[EV],MAXFN-EV) # replace NAs
    RES[[2]][,R]=F # store all best values
    VAL[R,2]=F[MAXFN] # store best value at MAXFN
  }

# compute average F result per method:
MIN=Inf
AV=matrix(nrow=MAXFN,ncol=length(Methods))
for(m in 1:length(Methods))
  for(i in 1:MAXFN)
   {
    AV[i,m]=mean(RES[[m]][i,])
    # update MIN for plot (different than -Inf):
    if(AV[i,m]!=-Inf && AV[i,m]<MIN) MIN=AV[i,m]
   }
# show results:
cat(Methods,"\n")
cat(round(apply(VAL,2,mean),digits=0)," (average best)\n")
# Mann-Whitney non-parametric test:
p=wilcox.test(VAL[,1],VAL[,2],paired=TRUE)$p.value
cat("p-value:",round(p,digits=2),"(<0.05)\n")

# create PDF file:
pdf("comp-bagprices-constr2.pdf",width=5,height=5,
    paper="special")
par(mar=c(4.0,4.0,1.8,0.6)) # reduce default plot margin
# use a grid to improve clarity:
g1=seq(1,MAXFN,length.out=500) # grid for lines
MAX=max(AV)
plot(g1,AV[g1,2],ylim=c(MIN,MAX),type="l",lwd=2,
     main="bag prices with constraint 2",
     ylab="average best",xlab="number of evaluations")
lines(g1,AV[g1,1],lwd=2,lty=2)
legend("bottomright",legend=rev(Methods),lwd=2,lty=1:4)
dev.off() # close the PDF device
```

5.4

```
library(genalg) # load rba.bin
library(GA) # load ga
library(mcga) # load mcga

library(parallel) # load detectCores() and others
library(tictoc) # load tic and toc

D=2
# common setup for the executions:
lower=rep(-512,D);upper=rep(512,D) # lower and upper bounds
Pop=400 # population size
Gen=500 # maximum number of generations
```

```
# eggholder function:
eggholder=function(x) # length of x is 2
  -(x[2]+47)*sin(sqrt(abs(x[2]+x[1]/2+47)))-x[1]*sin(sqrt(abs(x
      [1]-(x[2]+47))))
eggholder2=function(x) -eggholder(x) # maximization goal
K=1000
eggholder3=function(x) # penalizes x values outside the [lower,
    upper] range
{ res=eggholder(x) #
  # death penalty if x is outside the [lower,upper] range:
  minx=min(x)
  maxx=max(x)
  if(minx<lower[1]||maxx>upper[1]) res=K
  return(res)
}

Dig=2 # use 2 digits to show results
# simple function to show best solution and evaluation value:
showres=function(s,f,digits=Dig)
 { cat("best:",round(s,Dig),"f:",round(f,Dig),"\n") }

# rbga (minimization):
cat("rbga:\n")
tic()
ga1=rbga(lower,upper,popSize=Pop,iters=Gen,evalFunc=eggholder)
b=which.min(ga1$evaluations) # best individual
# show best solution and evaluation value:
showres(ga1$population[b,],ga1$evaluations[b])
toc()

# gaisl (maximization):
cat("gaisl:\n")
NC=detectCores()
tic()
cat("gaisl with ",NC,"cores:\n")
ga2=gaisl(type="real-valued",fitness=eggholder2,
          lower=lower,upper=upper, # lower and upper bounds
          popSize=Pop,
          numIslands=NC, # number of Islands
          parallel=TRUE,
          maxiter=Gen,monitor=FALSE)
toc()
showres(ga2@solution,-ga2@fitnessValue)

# mcga (minimization):
cat("mcga:\n") # uniform binary crossover and bit mutation
tic()
ga3=mcga(popsize=Pop,chsize=D,
         minval=lower,maxval=upper,
         maxiter=Gen,evalFunc=eggholder3)
toc()
showres(ga3$population[1,],ga3$costs[1])
```

5.5

```
# solution: s5-5-1.R
library(rgp) # load rgp

# auxiliary functions:
mae=function(y1,y2) mean(abs(y1-y2)) # mean absolute error
    function
eggholder=function(x) # length of x is 2
  -(x[2]+47)*sin(sqrt(abs(x[2]+x[1]/2+47)))-x[1]*sin(sqrt(abs(x
      [1]-(x[2]+47))))

fwrapper=function(x,f)
{ res=suppressWarnings(f(x[1],x[2]))
  # if NaN is generated (e.g. sqrt(-1)) then
  if(is.nan(res)) res=Inf # replace by Inf
  return(res)
}

# configuration of the genetic programming:
ST=inputVariableSet("x1","x2")
cF1=constantFactorySet(function() sample(c(2,47),1) )
FS=functionSet("+","-","/","sin","sqrt","abs")
# set the input samples:
samples=100
domain=matrix(ncol=2,nrow=samples)
domain[]=runif(samples,-512,512)
y=apply(domain,1,eggholder)
eval=function(f) # evaluation function
  mse(y,apply(domain,1,fwrapper,f))

# run the genetic programming:
gp=geneticProgramming(functionSet=FS,inputVariables=ST,
                      constantSet=cF1,populationSize=100,
                      fitnessFunction=eval,
                      stopCondition=makeStepsStopCondition(50),
                      verbose=TRUE)
# show the results:
b=gp$population[[which.min(gp$fitnessValues)]]
cat("best solution (f=",eval(b),"):\n")
print(b)
y2=apply(domain,1,fwrapper,b)
# sort L1 and L2 (according to L1 indexes)
# for an easier comparison of both curves:
MIN=min(y,y2);MAX=max(y,y2)
plot(y,ylim=c(MIN,MAX),type="l",lwd=2,lty=1,
     xlab="points",ylab="function values")
lines(y2,type="l",lwd=2,lty=2)
legend("bottomright",leg=c("eggholder","GP function"),lwd=2,lty
    =1:2)
# note: the fit is not perfect, but the search space is too large

# solution: s5-5-2.R
library(gramEvol) # load gramEvol
```

```
# auxiliary functions:
mae=function(y1,y2) mean(abs(y1-y2)) # mean absolute error
    function
eggholder=function(x) # length of x is 2
  -(x[2]+47)*sin(sqrt(abs(x[2]+x[1]/2+47)))-x[1]*sin(sqrt(abs(x
      [1]-(x[2]+47))))

# set the grammar rules:
ruleDef=list(expr=gsrule("<expr><op><expr2>","<func>(<expr>)","
    <expr2>"),
             op=gsrule("+","-","*"),
             func=gsrule("sin","sqrt","abs"),
             expr2=gsrule("x[1]","x[2]","<value>"),
             value=gsrule("<digits>.<digits>"),
             digits=gsrule("<digits><digit>","<digit>"),
             digit=grule(0,1,2,3,4,5,6,7,8,9)
            )
# create the BNF grammar object:
gDef=CreateGrammar(ruleDef)

# grammatical evolution setup:
Pop=100 # population size
Gen=50 # number of generations
# simple monitoring function:
monitor=function(results)
{ # print(str(results)) shows all results components
  iter=results$population$currentIteration # current iteration
  f=results$best$cost # best fitness value
  if(iter==1||iter%%10==0) # show 1st and every 10 iter
    cat("iter:",iter,"f:",f,"\n")
}

# set the input samples:
samples=100
domain=matrix(ncol=2,nrow=samples)
domain[]=runif(samples,-512,512)
y=apply(domain,1,eggholder) # compute target output

K=1000 # large penalty value
eval1=function(x,expr) # x is an input vector with D=2
{ # expr can include x[1] or x[2] symbols
  #print(expr)
  res=suppressWarnings(eval(expr)) # can generate NaNs
  if(is.nan(res)) res=K
  return(res)
}

maevalue=function(expr) # evaluation function
{
 y_expr=apply(domain,1,eval1,expr) # expr outputs for domain
 return (mae(y,y_expr))
}
```

```
set.seed(12345) # set for replicability
# run the grammar evolution:
ge=suppressWarnings(GrammaticalEvolution(gDef,maevalue,popSize=
    Pop,
                              iterations=Gen,monitorFunc=monitor))
b=ge$best # best solution
cat("evolved phenotype:")
print(b$expression)
cat("f:",b$cost,"\n")

# create approximation plot:
y2=apply(domain,1,eval1,b$expression)
MIN=min(y,L2);MAX=max(y,y2)
plot(y,ylim=c(MIN,MAX),type="l",lwd=2,lty=1,
     xlab="points",ylab="function values")
lines(y2,type="l",lwd=2,lty=2)
legend("bottomright",leg=c("eggholder","GE function"),lwd=2,
       lty=1:2)
```

Exercises of Chapter 6

6.1

```
source("hill.R") # load the blind search methods
source("mo-tasks.R") # load MO bag prices task
source("lg-ga.R") # load tournament function

# lexicographic hill climbing, assumes minimization goal:
lhclimbing=function(par,fn,change,lower,upper,control,
                    ...)
{
  for(i in 1:control$maxit)
     {
      par1=change(par,lower,upper)
      if(control$REPORT>0 &&(i==1||i%%control$REPORT==0))
         cat("i:",i,"s:",par,"f:",eval(par),"s'",par1,"f:",
             eval(par1),"\n")
      pop=rbind(par,par1) # population with 2 solutions
      I=tournament(pop,fn,k=2,n=1,m=2)
      par=pop[I,]
     }
  if(control$REPORT>=1) cat("best:",par,"f:",eval(par),"\n")
  return(list(sol=par,eval=eval(par)))
}

# lexico. hill climbing for all bag prices, one run:
D=5; C=list(maxit=10000,REPORT=10000) # 10000 iterations
s=sample(1:1000,D,replace=TRUE) # initial search
ichange=function(par,lower,upper) # integer value change
{ hchange(par,lower,upper,rnorm,mean=0,sd=1) }
LEXI=c(0.1,0.1) # explicitly defined lexico. tolerances
```

```
eval=function(x) c(-profit(x),produced(x))
b=lhclimbing(s,fn=eval,change=ichange,lower=rep(1,D),
          upper=rep(1000,D),control=C)
cat("final ",b$sol,"f(",profit(b$sol),",",produced(b$sol),")\n")
```

6.2

```
library(genalg) # load rbga function
library(mco)  # load nsga2 function

set.seed(12345) # set for replicability

# real value FES2 benchmark:
fes2=function(x)
{ D=length(x);f=rep(0,3)
  for(i in 1:D)
     {
      f[1]=f[1]+(x[i]-0.5*cos(10*pi/D)-0.5)^2
      f[2]=f[2]+abs(x[i]-(sin(i-1))^2*(cos(i-1)^2))^0.5
      f[3]=f[3]+abs(x[i]-0.25*cos(i-1)*cos(2*i-2)-0.5)^0.5
     }
  return(f)
}

D=8;m=3

# WBGA execution:
# evaluation function for WBGA
# (also used to print and get last population fes2 values:
# WBGA chromosome used: x=(w1,w2,w3,v1,v2,v3,...,vD)
#   where w_i are the weights and v_j the values
eval=function(x,REPORT=FALSE)
{ D=length(x)/2
  # normalize weights, such that sum(w)=1
  w=x[1:m]/sum(x[1:m]);v=x[(m+1):length(x)];f=fes2(v)
  if(REPORT)
    { cat("w:",round(w,2),"v:",round(v,2),"f:",round(f,2),"\n")
      return(f)
    }
  else return(sum(w*f))
}
WBGA=rbga(evalFunc=eval,
          stringMin=rep(0,D*2),stringMax=rep(1,D*2),
          popSize=20,iters=100)
print("WBGA last population:")
# S1 contains the Pareto curve: 20 solutions x 3 objectives
S1=t(apply(WBGA$population,1,eval,REPORT=TRUE))

# NSGA-II execution:
NSGA2=mco::nsga2(fn=fes2,idim=D,odim=m,
        lower.bounds=rep(0,D),upper.bounds=rep(1,D),
        popsize=20,generations=100)
# S2 contains the Pareto curve: 20 solutions x 3 objectives
S2=NSGA2$value[NSGA2$pareto.optimal,]
```

```
print("NSGA2 last Pareto front:")
print(round(S2,2))

# Comparison of results:
ref=c(2.0,8.0,10.0)
hv1=dominatedHypervolume(S1,ref)
cat("WGA hypervolume",round(hv1,2),"\n")
hv2=dominatedHypervolume(S2,ref)
cat("NSGA-II hypervolume",round(hv2,2),"\n")
```

Exercises of Chapter 7

7.1

```
# cycle crossover (CX) operator:
# m is a matrix with 2 parent x ordered solutions
cx=function(m)
{
 N=ncol(m)
 c=matrix(rep(NA,N*2),ncol=N)
 stop=FALSE
 k=1
 ALL=1:N
 while(length(ALL)>0)
 {
  i=ALL[1]
  # perform a cycle:
  base=m[1,i];vi=m[2,i]
  I=i
  while(vi!=base)
   {
    i=which(m[1,]==m[2,i])
    vi=m[2,i]
    I=c(I,i)
   }
  ALL=setdiff(ALL,I)
  if(k%%2==1) c[,I]=m[,I] else c[,I]=m[2:1,I]
  k=k+1
 }
 return(c)
}

# example of CX operator:
m=matrix(ncol=9,nrow=2)
m[1,]=1:9
m[2,]=c(9,8,1,2,3,4,5,6,7)
print(m)
print("---")
print(cx(m))
```

7.2

```
### code copied from tsf.R file:
# get sunspot yearly series:
url="http://www.sidc.be/silso/DATA/SN_y_tot_V2.0.txt"
series=read.table(url)
# lets consider data from the 1700-2019 years:
# lets consider column 2: sunspot numbers
series=series[1:320,2]
# save to CSV file, for posterior usage (if needed)
write.table(series,file="sunspots.csv",
            col.names=FALSE,row.names=FALSE)

L=length(series) # series length
forecasts=32 # number of 1-ahead forecasts
outsamples=series[(L-forecasts+1):L] # out-of-samples
sunspots=series[1:(L-forecasts)] # in-samples
LIN=length(sunspots) # length of in-samples

# mean absolute error of residuals
maeres=function(residuals) mean(abs(residuals))

INIT=10 # initialization period (no error computed before)
library(forecast) # load forecast package
arima=auto.arima(sunspots) # detected order is AR=2, MA=1
print(arima) # show ARIMA model
cat("arima fit MAE=",
    maeres(arima$residuals[INIT:length(sunspots)]),"\n")

### s7-2.R solution code:
library(pso) # load pso

# evaluation function of arma coefficients:
# s is a vector with 4 real values
# (ar1,ar2,ma1,intercept or m)
evalarma=function(s)
{ a=suppressWarnings(arima(sunspots,order=c(AR,0,MA),fixed=s))
  R=a$residuals[INIT:length(sunspots)]
  R=maeres(R)
  if(is.nan(R)) R=Inf # death penalty
  return(maeres(R))
}

AR=2;MA=1
maxit=50; LP=20
meants=mean(sunspots);K=0.1*meants
lower=c(rep(-1,(AR+MA)),meants-K)
upper=c(rep(1,(AR+MA)),meants+K)
C=list(maxit=maxit,s=LP,trace=10,REPORT=10)
set.seed(12345) # set for replicability
PSO=psoptim(rep(NA,length(lower)),fn=evalarma,
              lower=lower,upper=upper,control=C)
arima2=arima(sunspots,order=c(AR,0,MA),fixed=PSO$par)
print(arima2)
```

```
cat("pso fit MAE=",PSO$value,"\n")

# one-step ahead predictions:
f1=rep(NA,forecasts)
f2=rep(NA,forecasts)
for(h in 1:forecasts)
  { # execute arima with fixed coefficients but with more
      in-samples:
    # normal ARIMA forecasts:
    arima1=arima(series[1:(LIN+h-1)],order=arima$arma[c(1,3,2)],
        fixed=arima$coef)
    f1[h]=forecast(arima1,h=1)$mean[1]
    # PSO ARIMA forecasts:
    arima2=arima(series[1:(LIN+h-1)],order=arima2$arma[c(1,3,2)],
        fixed=arima2$coef)
    f2[h]=forecast(arima2,h=1)$mean[1]
  }
e1=maeres(outsamples-f1)
e2=maeres(outsamples-f2)
text1=paste("arima (MAE=",round(e1,digits=1),")",sep="")
text2=paste("pso arima (MAE=",round(e2,digits=1),")",sep="")

# show quality of one-step ahead forecasts:
ymin=min(c(outsamples,f1,f3))
ymax=max(c(outsamples,f1,f3))
plot(outsamples,ylim=c(ymin,ymax),type="l",
     xlab="time",ylab="values")
lines(f1,lty=2,type="b",pch=3,cex=0.5)
lines(f2,lty=3,type="b",pch=5,cex=0.5)
legend("topright",c("sunspots",text1,text2),lty=1:3,pch=c(1,3,5))
```

7.3

```
library(rminer) # load rminer package
library(mco) # load mco package

# load wine quality dataset directly from UCI repository:
file="http://archive.ics.uci.edu/ml/machine-learning-databases/
    wine-quality/winequality-white.csv"
d=read.table(file=file,sep=";",header=TRUE)

# convert the output variable into 2 classes of wine:
# "low" <- 3,4,5 or 6; "high" <- 7, 8 or 9
d$quality=cut(d$quality,c(1,6,10),c("low","high"))

# to speed up the demonstration, only 25% of the data is used:
n=nrow(d) # total number of samples
ns=round(n*0.25) # select a quarter of the samples
set.seed(12345) # for replicability
ALL=sample(1:n,ns) # contains 25% of the index samples
w=d[ALL,] # new wine quality data.frame
# show the attribute names:
cat("attributes:",names(w),"\n")
cat("output class distribution (25% samples):\n")
```

```
print(table(w$quality)) # show distribution of classes

# save dataset to a local CSV file:
write.table(w,"wq25.csv",col.names=TRUE,row.names=FALSE,sep=";")

# holdout split into training (70%) and test data (30%):
H=holdout(w$quality,ratio=0.7)
cat("nr. training samples:",length(H$tr),"\n")
cat("nr. test samples:",length(H$ts),"\n")
# save to file the holdout split index:
save(H,file="wine-H.txt",ascii=TRUE)

output=ncol(w) # output target index (last column)
maxinputs=output-1 # number of maximum inputs

# auxiliary functions:
# rescale x from [0,1] to [min,max] domain:
transf=function(x,min,max) return (x*(max-min)+min)
# decode the x genome into the model hyperparameters:
# Model is a global variable with "randomForest" or "ksvm"
decode=function(x)
{ if(Model=="randomForest")
   { # 2 hyperparameters:
     ntree=round(transf(x[1],1,200))
     mtry=round(transf(x[2],1,11))
     return(c(ntree,mtry))
   }
  else
   { # 2 SVM hyperparameters:
     sigma=transf(x[1],2^-6,2^1)
     C=transf(x[2],2^-2,2^5)
     return(c(sigma,C))
   }
}

# evaluation function (requires some computation):
# Model is a global variable with "randomForest" or "ksvm"
evalmodel=function(x) # x is a solution
{
 # read input features: from position 1 to maxinputs
 features=round(x[1:maxinputs]) # 0 or 1 vector
 inputs=which(features==1) # indexes with 1 values
 # use first feature if inputs is empty
 if(length(inputs)==0) inputs=1
 J=c(inputs,output) # attributes
 k3=c("kfold",3,123) # internal 3-fold validation
 # read hyperparameters:
 hpar=decode(x[(maxinputs+1):length(x)])
 if(Model=="randomForest")
   M=suppressWarnings(try(
       mining(quality~.,w[H$tr,J],method=k3,
              model="randomForest",ntree=hpar[1],
              # mtry cannot be higher than I - the
              #  number of inputs
```

```
            mtry=min(hpar[2],length(inputs))
           )
                     ,silent=TRUE))
 else M=suppressWarnings(try(
       mining(quality~.,w[H$tr,J],method=k3,
              model="ksvm",kpar=list(sigma=hpar[1]),C=hpar[2])
                     ,silent=TRUE))

 # AUC for the internal 3-fold cross-validation:
 if(class(M)=="try-error") auc=0.5 # worst auc
 else auc=as.numeric(mmetric(M,metric="AUC"))
 auc1=1-auc # maximization into minimization goal
 ninputs=length(inputs) # number of input features
 EVALS<<-EVALS+1 # update evaluations
 if(EVALS==1||EVALS%%Psize==0) # show current evaluation:
    cat(EVALS," evaluations (AUC: ",round(auc,2),
        " nr.features:",ninputs,")\n",sep="")
 return(c(auc1,ninputs)) # 1-auc,ninputs
}

# NSGAII multi-objective optimization:
m=2 # two objectives: AUC and number of input features
hrf=2 # number of hyperparameters for randomForest
genome=maxinputs+hrf # genome length
lower=rep(0,genome)
upper=rep(1,genome)
EVALS<<-0 # global variable
PTM=proc.time() # start clock
Psize=20 # population size

set.seed(12345) # for replicability
Model="randomForest"
cat("nsga2 for ",Model,":\n")
s1=mco::nsga2(fn=evalmodel,idim=length(lower),odim=m,
         lower.bounds=lower,upper.bounds=upper,
         popsize=Psize,generations=10)
sec=(proc.time()-PTM)[3] # get seconds elapsed
cat("time elapsed:",sec,"\n")

set.seed(12345) # for replicability
EVALS<<-0 # global variable
Model="ksvm"
cat("nsga2 for ",Model,":\n")
s2=mco::nsga2(fn=evalmodel,idim=length(lower),odim=m,
         lower.bounds=lower,upper.bounds=upper,
         popsize=Psize,generations=10)
sec=(proc.time()-PTM)[3] # get seconds elapsed
cat("time elapsed:",sec,"\n")

# plot the Pareto fronts:
par(mfrow=c(1,2))  # plotting area with 1*2 array
# 1st plot
po1=which(s1$pareto.optimal) # optimal points
# sort Pareto front according to f2 (number of features):
```

```
i1=sort.int(s1$value[po1,2],index.return=TRUE)
pareto1=s1$value[po1[i1$ix],] # Pareto front f1,f2 values
plot(1-pareto1[,1],pareto1[,2],xlab="AUC",
     ylab="nr. features",type="b",lwd=2,main="randomForest Pareto
         :")
# 2nd plot
po2=which(s2$pareto.optimal) # optimal points
# sort Pareto front according to f2 (number of features):
i2=sort.int(s2$value[po2,2],index.return=TRUE)
pareto2=s2$value[po2[i2$ix],] # Pareto front f1,f2 values
plot(1-pareto2[,1],pareto2[,2],xlab="AUC",
     ylab="nr. features",type="b",lwd=2,main="ksvm Pareto:")
```

Index

P. Cortez, *Modern Optimization with R*, Use R!,
https://doi.org/10.1007/978-3-030-72819-9

Printed in the United States
by Baker & Taylor Publisher Services

Printed in the United States
by Baker & Taylor Publisher Services